Scie

J J Wellington

The *Questions* Publishing Company Ltd
Birmingham

s Publishing Company Ltd
irmingham B1 3HH

4

Designed by Iqbal Aslam
Illustrations by Martin Lealan and Trevor Carter

Printed in Great Britain

Acknowledgements

I would like to thank several people for helping me to prepare this Glossary of Science Words: first of all, Val Ross and John Corbett of Manor Lodge Junior School, Sheffield for getting me started and providing valuable feedback at the beginning; secondly, Harmony and Trudi Esberger, Ruth Lawless, Tilly Ross and Wendy Wellington for their encouragement; my science colleague, Jenny Henderson for checking the accuracy and readability of the entries; and last but not least, Hannah Wellington for going through every single entry, making corrections and suggestions, and checking that they all make sense.

I hope that pupils, parents and teachers find this Dictionary helpful in learning science.

J J Wellington

The Science Dictionary

Learning science involves learning its language. Many studies have shown that the language barrier is one of the biggest barriers to learning science. This Dictionary has been written in order to help learners, **of all ages**, to help tackle that barrier.

Every teacher and every pupil learning science will be able to make good use of it – so will their parents.

What's in the dictionary?

The Dictionary covers all the key words and terms of the National Curriculum of England and Wales, and large parts of the science curriculum in other countries such as: Scotland, Australia and the USA. The words to be included have been carefully selected by examining curriculum documents and by consulting over 30 teachers to decide on the final list.

Each word is explained in full (not just defined) and illustrations are used to help give it meaning and sense. Many of the words are related to each other.

It is designed to be helpful both for pupils (mainly at ages 7 to 16) **and** teachers.

Who can use it . . . and how?

The Dictionary has been tried and tested with teachers and with pupils. The trials show that it will be valuable for both, and parents will also find it very useful.

For teaching . . .

Teachers can use the Dictionary to:

- highlight new words which will occur in teaching a topic, e.g. electricity, energy, food, water, plant growth . . . The words could be singled out, photocopied and made into a poster.

- many teachers are teaching outside their own subject specialism and it will be useful to them if they need 'refreshment', reminder or in some cases if the word is completely new to them. A bit of pre-lesson revision is always useful (even in our **own** specialism) and also helps to remind us of the importance of language in science teaching.

For learning...

Pupils can use the Dictionary:

- for revising or simply refreshing their memory.

- when writing about science, e.g. a story, a description, an account of an investigation, a write-up of an experiment. The Dictionary will help them to use words accurately and to stimulate new ideas and vocabulary to use in writing.

- in reading about science, e.g. a science textbook, a story about science, a newspaper article or a piece in a magazine. The Dictionary can help readers to understand the writing, to check its accuracy and to look for other words and ideas which connect with it (these are highlighted in **bold**).

- in discussing or just talking about science, e.g. to clarify words; to look for new words, or to connect words and ideas together.

For parents...

Parents, especially if they have not studied science for some time, will find the Dictionary helpful in:

- refreshing their memory.

- learning some of the newer words and ideas of science.

- as a general reference.

It may even help them to help their children doing science homework.

The author

Jerry Wellington taught science in Tower Hamlets, East London before joining the University of Sheffield as a lecturer in science education. He has written many science textbooks for schools and homes including: *Sounds, Energy*, and *Physics for All* (Stanley Thornes); *Beginning Science: Physics* (Oxford University Press); *The Super Science Book of Space* (Wayland).

He also writes regularly for the *Times Educational Supplement* and has produced books for teachers such as: *Secondary Science: contemporary issues and practical approaches* and *Practical Work in Science: which way now?* (both London: Routledge).

Using the symbols

The language of science comes in many shapes and forms: naming words, abstract ideas and concepts, names of processes (some artificial, some natural). The 3 symbols below have been used to show some of the different types of words used in science. Each entry in the Dictionary has one of these symbols next to it:

Symbols

Naming words

New names for familiar objects or things, e.g. urine
Names specific to chemicals, physical objects or living things, e.g. vertebrate
Names of units used in science, e.g. joule, watt

Process words

Natural processes, e.g. evaporation, photosynthesis
Artificial processes, e.g. fusion, fission

Idea words

Ideas with everyday **and** scientific meanings, e.g. energy, power
Ideas that are scientific concepts only, e.g. atoms, particles

Notes

Note headings in **Bold** refer to dictionary entries. Note headings in *Italic* are commonly used phrases and words which are cross-referenced to entries in the dictionary.

A

Acceleration
Children can be asked to consider the car, skateboarding and diving examples and then come up with some examples of their own: stone falling off a cliff, pedal cyclist going down hill, sprinter leaving the starting blocks, etc. Can they give the force, or forces, at work in each of their examples?

Acid
The acid in vinegar is acetic acid (from the Latin acetum = vinegar), that in citrus fruits is citric acid, that in an ant's sting is formic acid (from the Latin formica = ant). Acids commonly used in laboratories are nitric acid, hydrochloric acid and sulphuric acid. Do children know the names of any acids?

Air
See: **Atmosphere, Friction** and **Gases**

Alcohol
See: **Expansion, Fermentation, Fungus** and **Glucose**

Alkali
These can be just as dangerous as acids. Where can the children spot the warning symbol for corrosive liquids? (Tankers, tanks, bottles, drums)

Aluminium
Can the children think of any more uses of aluminium? (Window frames, milk bottle tops, sardine tins, Christmas decorations)

Ampere
See: **Ohm** and **Current**

Amplitude
It should be understood that amplitude is a measure of the *size* of a vibration, *not how quickly* the vibrating object goes to and fro.

Animal
Can the children say how animals are different to plants?
　See also: **Reproduction**

Antibiotic
It should be noted that antibiotics are not effective against viruses.

Artery
Children should find out about differences between the blood in the arteries and the blood in the veins.
   See also: **Blood**

Atmosphere
Can the children think where they can see water vapour in the atmosphere? (Clouds, mist, fog, breathing out on a cold day, trails left in the sky by jets)
   See also: **Photosynthesis, Barometer** and **Pressure**

Atom
See also: **Electron, Element, Fission, Fusion, Gamma rays, Gases, Particle** and **Periodic Table**

B

Bacterium
See: **Antibiotic, Microbe** and **Virus**

Barometer
Why would water *not* be a very good liquid for a barometer? Why is mercury better? Why are mercury barometers *not* very good for carrying around? Why do you think a barometer is called an altimeter when it is used to measure the height above sea-level of something, e.g. a plane?

Biodegradable
Can you think of some things that are biodegradable? Which things or materials have you seen that are not biodegradable? No litter is good, but why do these materials cause a worse litter problem than, say, apple cores or orange peel?
   See also: **Decompose** and **Microbe**

Blood
Teachers could discuss *blood groups* and *transfusion* with children. Do they know their own blood groups? Do they know anyone who donates blood or has had a transfusion? Have any of the children had a blood transfusion? What are the dangers of transfusion? Teachers might also discuss *immunity* and *vaccination/inoculation*. What 'jabs' have pupils had? What are *vaccines* and why are they useful?

   Other topics which could be discussed include *haemophilia* (blood clotting slowly or not at all); the effect of cutting a main artery; blood disorders and so on. Children seem to have no shortage of blood-related anecdotes.

Boiling
See: **Celsius, Liquid, Melt** and **Solid**

C

Camera
Why is the eye like a lens camera? How is it different?
   Pupils may know that the image on the film of a camera and the retina of the eye is actually *upside down*. Our brain turns it 'the right way up'.
   See also: **Image**

Capillary
See also: **Artery, Blood, Carbon dioxide, Heart**

Carbohydrate
Pupils can see this word and compare the amounts in different foods by looking at labels on packets (cereals, crisps, etc) and cans (soft drinks).

They may already know that carbohydrates are present in potatoes, bread, pasta etc, that they give us energy, and that they can make us fat!

Carbon
The soot that comes from a candle flame is carbon. You can collect some soot for examination by holding an object (e.g. an old spoon) above a burning candle.

Carbon dioxide
Most pupils will have heard of carbon dioxide in connection with the *greenhouse effect* (see later entry) and the problems of global warming. It would be interesting to ask them if/where they have heard of carbon dioxide before.

Carnivore
Discussion of this word could lead to an interesting debate on why many animals, including humans, eat meat. Why do some people not eat meat? How do they survive? Can pupils think of some other carnivores and **herbivores**?

Catalyst
See: **Enzyme** and **Reaction**

Cell
Teachers can discuss with children the various types of body tissue and how each has its own type of cells. If nerves have nerve cells, what type of cells does skin have? And muscles? And bones? What other examples can the children think of?

By looking at a plant that has not had enough water, children can see how water is vital to keep up the supply of cell sap in the vacuoles of the plant cells. When the sap is not pressing against the cell walls, the plant wilts.

Chlorine
Discussion of the use of chlorine to treat water can be linked to children's work in history on the growth of towns and public health.

Chlorophyll
Children can think about what season it is when many trees lose their green colour, and what happens to the trees then.
See also: **Carbon dioxide** and **Cell**

Chromosome
See: **Genes**

Circuit
Pupils will almost certainly have heard this word in other contexts, particularly motor racing, and this should help them to understand the key idea that electricity needs a complete path in order to travel at all.
See also: **Conduction** and **Current**

Combustion
See also: **Fuel, Oxygen, Reaction** and **Respiration**

Compounds
An interesting example of the way that compounds differ from the elements that form them is salt (sodium chloride). Salt is completely different from sodium, which is a grey metal that has to be stored in oil because with water it sets free hydrogen (which is flammable) and a lot of heat, and from chlorine, which is a green, poisonous gas.
   See also: **Element**

Compression
Children can try most of the examples suggested here for themselves. The pump and syringe examples are relevant to technology work on pneumatics and hydraulics.

Concave
See also: **Lens, Reflection** and **Refraction**

Conduction
See also: **Circuit** and **Insulation**

Constellation
It should be pointed out that although the stars in a constellation may look as though they form a flat pattern, some of them may well be millions of miles behind, or in front of, others.

Convection
Children can try (with parental supervision) producing an example of convection themselves by running a bath without mixing the water. This often produces a hot layer of water on top, and a cold layer at the bottom.

Convex
See also: **Lens, Reflection** and **Refraction**

Crystal
The easiest crystals for children to examine are salt and sugar. However, more exciting (and beautiful) for children to consider are the crystals that are precious and semi-precious stones.

Current
Common non-conductors include plastic, rubber and wool.

D

Decibel
See: **Sound**

Decompose
Children can look for products which have the description *bio-degradable* on them, for example in supermarkets. They can investigate 'bio-degradable' alternatives to materials that are used in everyday products.
   Charcoal is a good example of the result of chemical compounds decomposing, as children are familiar with it for drawing and as a fuel for barbecues.

Diaphragm
See also: **Lung**

Diffusion
See also: **Gases, Helium,** and **Hydrogen**

Digestion
See also: **Enzyme** and **Excretion**

E

Echo
Echoes can also be used to find the depth of the sea bed underneath a ship. A wave is sent out from the bottom of the ship, bounces off the sea bed, and comes back to the ship. The depth of the sea can be found if we know the time taken for the echo to come back and also the speed of sound in water.

For example, suppose the wave is sent out and the echo returns two seconds later. If the speed of sound in water is 1,500 metres per second the sound wave must have travelled 1,500 x 2 or 3,000 metres. It has to get there and back. So the sea must be 3,000/2 or 1,500 metres deep.

Ecosystem
See also: **Habitat** and **Organism**

Electricity
See: **Circuit, Current** and **Energy**

Electromagnet
See also: **Current** and **Magnet**

Electromagnetic spectrum
See: **Prism, Radiation, Reflection, Ultraviolet** and **Vacuum**

Electron
See also: **Atom, Current** and **Nucleus**

Element
See also: **Atom, Compounds** and **Molecule**

Embryo
See also: **Fertilisation, Foetus, Gamete, Reproduction, Sperm** and **Uterus**

Energy
Kinetic energy is named after the Greek word for movement, *kinesis*.
See also: **Light** and **Sound**

Enzyme
See also: **Catalyst, Digestion, Protein, Reaction** and **Respiration**

Equilibrium
Related words: stable, unstable, centre of gravity. Can pupils think of other things (or even

people) with a low centre of gravity? Or even other objects, like table lamps, which are in stable equilibrium?

Evaporation
If you are wet your body is cooled as the water on your skin evaporates. This is called 'cooling by evaporation'. Inside a fridge, there is a special liquid inside pipes which evaporates and takes heat away from the inside of the fridge, just as drying sweat takes heat from your body. Outside the fridge cabinet the gas in the pipe is made to condense again, and gives out heat. That is why it often feels hot near the outside of a fridge.

Evolution
See also: **Extinction**

Excretion
The removal of undigested food from the body through the bowels is called **egestion**.
   See also: **Artery, Kidney** and **Digestion**

Expansion
Railway lines expand when they get hot. Years ago, gaps were left between the rails to allow for expansion, otherwise the rails would buckle in the heat. Nowadays, long rails are linked by overlapping joints.
   See also: **Alcohol**

Extinction
See also: **Evolution**

Eye
See: **Camera, Image, Lens** and **Ligaments**

F

Fermentation
See also: **Alcohol, Carbon dioxide, Enzyme** and **Glucose**

Fertilisers
Fertilisers can cause pollution if the minerals from them are washed away into rivers and lakes and enter our drinking water. Modern fertilisers are based mainly upon a chemical called *ammonia*.
   See also: **Decompose, Mineral** and **Pollution**

Fertilization
See also: **Embryo, Gamete, Ovary, Ovum, Reproduction** and **Sperm**

Fission
See also: **Atom** and **Nucleus**

Foetus
See also: **Embryo, Placenta** and **Uterus**

Food chain
See: **Carnivore, Ecosystem** and **Omnivore**

Food web
See: **Carnivore** and **Ecosystem**

Force
See: **Newton, Pressure** and **Weight**

Fossils
Coal and oil are fossilised organisms which we now call *fossil fuels*.
   See also: **Evolution, Organism** and **Pollution**

Frequency
See also: **Amplitude, Echo** (which bats use to find their way around), and **Waves**

Friction
'Shooting stars' are rocks from space which burn up when they enter the Earth's atmosphere, because friction between the air and very fast-moving objects makes large quantities of heat.
   See also: **Atmosphere**

Fuel
See also: **Carbon, Carbon dioxide** and **Combustion**

Fungus
See also: **Alcohol, Carbon dioxide, Cells, Decompose, Fermentation,** and **Organism**

Fuse
The word 'fuse' can also mean 'join together'. In this sense, cells and the nuclei of atoms can be fused. In the context of electricity, it means to liquefy or melt, or to join together by melting.
   See also: **Circuit** and **Current**

Fusion
See also: **Atom, Energy, Fission** and **Nucleus**

G

Galaxy
The Galaxy in which the earth is located is a collection of approximately 100,000 million stars. Our sun is located about two-thirds of the way from the centre of the galaxy.

Gamete
See also: **Cell, Embryo, Ovum, Sperm,** and **Reproduction**

Gamma rays
See also: **Atom, Frequency, Nucleus** and **Radiation**

Gases
See also: **Air, Atmosphere, Atoms** and **Chlorine**

Generators
See also: **Current** and **Magnet**

Genes
See also: **Chromosome, Fertilisation, Ovary** and **Sperm**

Germination
Pupils can carry out a simple investigation to see which conditions are needed for seeds to germinate, e.g. compare warm, dry sand or soil with ice-cold, dry sand or soil, or warm, wet sand or soil with cold, wet sand or soil.

Glucose
See also: **Alcohol, Carbohydrate, Energy, Enzyme, Fermentation** and **Photosynthesis**

Gravity
One newton (1N) is the force which will accelerate 1kg of mass by 1 metre per second per second.

Greenhouse effect
See: **Atmosphere, Carbon dioxide, Fuel** and **Renewable**

H

Habitat
Pupils can discuss the habitat they live in and what it is like (urban/rural; hostile/friendly; hot/cool; dry/wet. How have they adapted to the habitat they live in? A community or habitat can be created, e.g. a small pond or aquarium, or a rotting log in a transparent tank.
   The word *ecology* means 'the study of ecosystems'. All of the world's ecosystems together form a balanced system called the *biosphere*.
   See also: **Ecosystem** and **Evolution**

Heart
In some vertebrate animals such as most reptiles, amphibians and fish, the heart does not have two pumps.
   See also: **Artery, Blood** and **Capillary**

Heat
See: **Energy** and **Infrared**

Helium
Helium is named after *helios*, the Greek for Sun.

Herbivore
See also: **Carnivore**

Hormone
The rapid growth of the growing tip of a shoot is controlled by plant growth hormones.

Hydrogen
See also: **Acid, Helium** and **Oxygen**

I

Igneous
See: **Magma, Metamorphosis** and **Rocks**

Image
See also: **Lens, Reflection** and **Refraction**

Immunity
A famous doctor named Edward Jenner (1749-1823) once injected a boy called James Phipps with cowpox, a disease of cattle. James caught cowpox but quickly recovered from it. Jenner then injected James with a similar, but much more dangerous, disease germ called smallpox. Luckily for James, the antibodies which his blood had made to fight off the cowpox protected him, which proved that vaccination (named after the Latin word for cow, *vacca*), really works.
See also: **Virus**

Indicator
The term *pH* comes from the German word *potenz* meaning 'power' and H, the symbol for hydrogen.
See also: **Acid** and **Alkali**

Infrared
See also: **Gamma rays, Light, Radiation, Ultraviolet** and **Vacuum**

Inheritance
See: **Genes**

Insulation
Electrical insulators are the opposite of conductors. They do not carry electricity and often 'hold' static electricity if they have been rubbed, e.g. a plastic comb.
Heat and sound insulation hinder or prevent the free flow of heat and sound.
See also: **Circuit, Conduction** and **Electron**

Invertebrate
See also: **Animal** and **Vertebrate**

J

Joule
The joule is named after a British scientist, James Prescott Joule who was born in Salford in 1818. He discovered that heat is a form of energy.
Note the use of *joule* with a lower case 'j' for the unit, but 'J' (upper case) when abbreviated. All the SI units follow this rule, e.g. *newton* and 'N' for short.
The unit *calorie* is often used in magazines and elsewhere to measure food values. Children (and teachers) should be encouraged to think in terms of kilojoules. Fortunately, food packets, e.g. cereals, now often give values in kJ.
During the course of a day different people need or use different amounts of energy.
A coal miner will need about 15,000 kilojoules in a day, while a baby might only require 4,000 kJ.
See also: **Energy, Newton** and **Weight**

K

Kidney
See also: **Excretion, Organ** and **Urine**

Kilogram
See also: **Mass** and **Weight**

    Mass and **weight** are difficult concepts and can be easily confused, especially as they are used interchangeably in everyday situations and sometimes (wrongly) in science. Note that mass is measured in kg and weight in newtons. Weight measures the pull of gravity on the object, so it will be greater down a mine and slightly less at the top of a mountain.
    Mass measures the 'amount of stuff' in the object, so it does not change wherever it is measured.

Kinetic
Kinetic energy can be changed into other forms of energy. If you rub your hand up and down your sleeve, friction causes the kinetic energy to change to heat energy. The kinetic energy in moving water can be used to turn a water turbine and generate electricity. Modern cars are designed to absorb kinetic energy by 'crumpling up' on impact.
All matter is made up of constantly moving molecules, which move faster as the temperature increases. At absolute zero (-273°C or 0 kelvin) they would stop moving.

L

Larva
The word *larva* dates from around 1650. It meant 'a ghost, hobgoblin or spectre'.

Lens
See also: **Concave** and **Convex**

Lever
A long lever will move a very heavy weight with very little effort. Archimedes is reputed to have said: 'Give me somewhere to stand, and I will move the earth'. The human body has levers: think of the arm (with the elbow as the pivot) or the leg (with the knee as the pivot).

Ligaments
Ligaments are stretchy, i.e. they are elastic. Tendons are not stretchy, i.e. they are inelastic.
    See also: **Ovary** and **Uterus**

Light
See: **Energy** and **Infrared**

Liquid
See also: **Magma, Particle** and **Solid**

Liver
See: **Organ**

Luminous
The idea that some objects are luminous and some are not is very important. Children may

believe that we see things because our eyes send out light rays.

They may not be aware that some apparently luminous objects, such as mirrors and the Moon, are reflectors of light.

Loudness
See also: **Amplitude** and **Frequency**

Lung
See also: **Carbon dioxide, Diaphragm, Respiration** and **Trachea** (windpipe)

Lymph
See also: **Blood, Capillary** and **Cell**

M

Magma
See also: **Gases, Liquid, Metamorphosis,** and **Solid**

Magnet
See also: **Electromagnet**

Magnetic field
See: **Current, Generators** and **Magnet**

Mammal
See also: **Organisms**

Mass
A large mass is much harder to get moving or accelerate than a small one. The same force will give a smaller mass a much greater acceleration than a large one.
See also: **Acceleration, Kilogram** and **Weight**

Menstruation
See also: **Hormone, Ovary** and **Uterus**

Melt
For tin the melting point is about 230°C, for lead it is about 327°C. Gold melts at a much higher temperature of 1063°C.

Metamorphosis
See also: **Larva** and **Magma**

Microbe
See also: **Virus**

Microwave
See also: **Satellite**

Mineral
See also: **Blood, Fertiliser** and **Nitrogen**

Mirror
See: **Image**

Molecule
See: **Element, Particle** and **Protein**

Moment
See also: **Lever** and **Pivot**

Muscle
Some muscles in your body (such as in your arm or leg) work when you deliberately control them, i.e. from your will-power. These are called voluntary. Other muscles are involuntary – your gut, heart and eye muscles should all work without your control.
   See also: **Heart** and **Ligaments**

N

Newton
See also: **Mass** and **Weight**

Nitrogen
Nitrogen turns to a liquid when it is very cold (-196°C). Liquid nitrogen in special flasks can be very useful for fast freezing and for keeping things refrigerated.
   See also: **Nutrition** and **Protein**

Nucleus
See also: **Cell, Electron** and **Hydrogen**

Nutrition
See also: **Carbohydrate, Energy, Mineral, Parasite, Protein,** and **Vitamin**

#

Ohm
The ohm is named after the German scientist Georg Ohm (1787-1854). He first stated the law that the larger the voltage in a circuit, the bigger the current.
   See also: **Current** and **Voltage**

Omnivore
See also: **Carnivore** and **Habitat**

Orbit
In the human body, the round socket in your skull for your eyeball is called the orbit.
   See also: **Atom, Electron, Microwave** and **Nucleus**

Organ
See also: **Heart, Lung, Kidney, Liver** and **Blood**

Organism
See also: **Mammal** and **Reptile**

Ovary
See also: **Fertilisation, Menstruation, Organ,** and **Uterus**

Ovum
See: **Fertilisation, Gamete, Genes** and **Ovary**

Oxygen
Many substances combine with oxygen to make a new substance called an oxide. When iron rusts it combines with oxygen from the air to make iron oxide, Copper makes copper oxide, carbon combines with oxygen to make carbon dioxide, sulphur makes a choking gas called sulphur dioxide. When a substance combines with oxygen and makes an oxide this is called oxidation.
See also: **Combustion, Energy, Glucose, Photosynthesis** and **Respiration**

Ozone
See also: **Atmosphere** and **Ultraviolet**

P

Parallel
See also: **Circuit**

Parasite
See also: **Organism**

Particle
See also: **Atom, Gases, Liquid, Melt, Molecule,** and **Solid**

Periodic table
See also: **Atom, Element** and **Nucleus**

Photosynthesis
See also: **Energy, Fossils, Fuels, Oxygen** and **Glucose**

Pitch
See: **Frequency, Sound** and **Ultrasonic**

Pivot
See also: **Lever, Moment** and **Muscle**

Placenta
Plants have a placenta – this is part of the inside wall of the ovary.
See also: **Embryo, Foetus** and **Uterus**

Planet
Asteroids are large rocks which orbit the Sun – they are like planets but much smaller. There is a belt of about 40,000 rocky asteroids between Mars and Jupiter. Comets are giant lumps of ice and rock which also orbit the Sun. They have very long orbits that go well outside the edge of the solar system.
See also: **Orbit**

Plants
See: **Animal, Fungus** and **Organism**

Pollen
See also: **Fertilisation** and **Gamete**

Pollution
Noise from traffic, aeroplanes, and people is a form of pollution. Unwanted radiation, such as radioactive waste, can be another form.
   See also: **Carbon dioxide, Fossils, Fuel** and **Radiation**

Power
See also: **Watt**

Pressure
See also: **Atmosphere, Barometer,** and **Vacuum**

Prism
See also: **Reflection, Refraction** and **Electromagnet**

Protein
Hormones and enzymes are proteins.
   See also: **Enzyme, Hormone, Molecules** and **Nutrition**

Pulse
See also: **Artery, Heart** and **Star**

R

Radiation
See also: **Gamma rays, Infrared** and **Pollution**

Reaction
See also: **Catalyst, Combustion, Compound, Periodic Table** and **Respiration**

Reflection
See also: **Concave, Convex, Echo** and **Luminous**

Refraction
See also: **Concave, Convex, Lens** and **Wave**

Renewable
Non-renewable sources are used up when we convert them into other forms of energy. Fuels like coal and oil are non-renewable. When coal is burnt it changes into mostly carbon dioxide which goes into the air.
   See also: **Energy, Fission, Fuel,** and **Nucleus**

Reproduction
See also: **Cell, Embryo, Fertilisation, Fission, Gamete, Placenta** and **Pollen**

Reptile
See also: **Organism** and **Vertebrate**

Resistance
Resistors are used inside electrical things because they get very hot when a current travels through them. Two examples are the heating coil or element in a hair drier, and the filament in a light bulb.
See also: **Ohm, Parallel** and **Series**

Respiration
See also: **Cell, Combustion, Fuel, Lung, Nutrition, Oxygen** and **Photosynthesis**

Rocks
The rocks of the Earth go round in a never ending cycle: the rock cycle. Igneous rocks are worn down to make sediments – these are deposited and squashed to make sedimentary rocks. Heat and pressure can change these to metamorphic rocks. Then metamorphic rocks can melt and turn back into *igneous* rocks.
See also: **Magma** and **Metamorphosis**

S

Satellite
See also: **Microwave** and **Orbit**

Seasons
See also: **Orbit**

Series
See also: **Resistance**

Skeleton
See also: **Ligament, Muscles, Reptile** and **Vertebrate**

Solid
See also: **Boiling, Evaporation, Gases, Liquid, Melt** and **Particle**

Soluble
To separate the solute, e.g. salt, from the solution you need to heat it in an evaporating dish. The solvent, e.g. water, evaporates into the air and the solute is left behind on the side of the dish.

Sound
See also: **Amplitude, Decibel, Echo, Frequency, Loudness, Reflection, Ultrasonic** and **Waves**

Sperm
See also: **Embryo, Fertilisation, Gamete** and **Ovary**

Star
See also: **Fusion, Luminous** and **Nucleus**

T

Teeth
Different animals have different sets of teeth, depending on what they eat. A dog and a lion have large canines for killing prey and holding it in their mouth. A sheep has sharp incisors for cutting grass and a good set of molars for chewing it. A shark has jagged teeth to cut through flesh.

Temperature
See also: **Blood, Capillary, Evaporation, Melt** and **Particles**

Tendon
See also: **Ligament, Muscle** and **Skeleton**

Thermometer
See: **Expansion** and **Temperature**

Tissue
See: **Cell** and **Lung**

Trachea
See: **Lung** and **Respiration**

U

Ultrasonic
See also: **Echo, Frequency** and **Sound**

Ultraviolet
See also: **Frequency, Ozone, Pollution** and **Wave**

Urine
See also: **Blood** and **Kidney**

Uterus
See also: **Embryo, Fertilisation, Ovary** and **Placenta**

V

Vaccination
See: **Antibiotic** and **Immunity**

Vacuum
See also: **Atmosphere, Barometer, Pressure, Sound** and **Waves**

Vein
See: **Artery, Blood, Capillary** and **Excretion**

Velocity
See: **Speed**

Vertebrate
See also: **Invertebrate, Reptile** and **Skeleton**

Vibration
See: **Frequency** and **Sound**

Virus
See also: **Antibiotic, Immunity** and **Microbe**

Vitamin
See also: **Minerals, Nutrition** and **Protein**

Voltage
See: **Circuit, Current, Ohm** and **Resistance**

W

Water cycle
See also: **Evaporation** and **Respiration**

Watt
See also: **Joule** and **Power**

Wavelength
See: **Amplitude, Frequency, Sound** and **Waves**

Waves
See also: **Amplitude, Frequency** and **Sound**

Weight
See also: **Mass, Newton** and **Force**

Womb
See: **Ovary** and **Uterus**

Work
See: **Joule, Power** and **Watt**

Acceleration

If you are driving and you press the **accelerator** pedal, the car picks up speed - it **accelerates**. If you press the brake pedal, the car slows down - it **decelerates**.

If you jump off a diving board, you **accelerate**.
If you roll down a hill on a skateboard, you **accelerate**.

A **force** is needed to make something **accelerate**. The **force** making you go faster as you dive or skateboard down a hill is **gravity**. In both cases, you will keep going faster until the force of **gravity** acting on you is **balanced** by the **forces** of **friction** and **air resistance**. At this point you will stop **accelerating** (but you will not slow down.)

The force making a car **accelerate** comes from the car's engine.

Acid

Ants have **acid** inside their stings. So do stinging nettles.

There are different kinds of **acid** in many of our foods.
Vinegar is a very weak **acid**.
The sharp taste of fruits like lemons and oranges comes from the **acid** in them.

If you drop some **metals**, e.g. **zinc**, into a strong **acid**, the **metal dissolves** and a **gas** called **hydrogen** is produced.

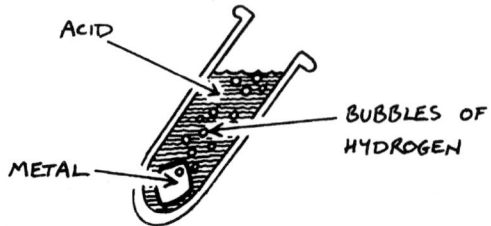

Strong **acids** are dangerous and must be handled with care. This is the warning symbol:

Acid (continued)

Acids have many uses.
Acid is made in your **stomach** to help **digest** (break down) food.
Acids can kill the **bacteria** in food, and so mild acids (like vinegar) are used to preserve things, e.g. pickled onions.

A **solution** is either **acid**, **alkaline** or **neutral** (neither **acid** nor **alkaline**). If a **solution** is **acid**, it will turn some blue vegetable dyes red. The redder the vegetable dye becomes, the stronger the **acid** is.

Alcohol

Alcohol is a liquid. Most people have heard of it because it is in **alcoholic** drinks. About one-twentieth of the liquid in beer is **alcohol**, and nearly half of whisky is **alcohol**.

Drinking too much **alcohol** affects people's brains. It can make them dizzy and stop them from seeing or thinking clearly.

Alcohol has many uses.
It kills **germs**, so it can be used to clean wounds.
When it burns, it provides energy, and car engines can be made that run on **alcohol**.

Alcohol is made from **plant** material (e.g. from **fruits** like grapes). It is made by letting the **sugar** in the **plant** material **ferment**. The **fermented** fruit mixture contains **alcohol** which can be **distilled** to make it even stronger.

Alkali

A **solution** of a **chemical** in water may be **acid** (e.g. lemon juice), **neutral** (e.g. pure water) or **alkaline** (e.g. ammonia cleaning fluid).

Substances can be tested using an **indicator** - usually a plant dye, such as **litmus**. If the substance is **alkaline**, it will turn red plant dye blue. If the substance is **acid**, it will turn blue plant dye red. The stronger the **alkali** or **acid**, the deeper the blue or red the dye turns.

The **acidity** or **alkalinity** of something is measured on the **pH** scale, which goes from 1 to 14. On this scale, **neutral** solutions measure 7. **Acid** solutions have a number less than 7. **Alkaline** solutions have a number greater than 7.

| 1 | 2 | 3 | 4 | 5 | 6 | 7 | 8 | 9 | 10 | 11 | 12 | 13 | 14 |

Alkali (continued)

When some **metals** are put into water, they **react** very strongly and form an **alkali**. These metals are called **alkali metals**. Two examples of such **metals** are **sodium** and **potassium**.

Strong **alkalis** are dangerous, for example to eyes and skin. This is the warning symbol:

Alkalis have various uses. The **alkali** in **ammonia** cleaning fluid helps get things like dirty ovens clean. The **alkali, calcium hydroxide**, in the lime that gardeners and farmers use **neutralises** soil that has too much **acid** in it.

Aluminium

As you toss the **aluminium** can from a soft drink into the recycling bank, you will notice how light it is.

Aluminium, which is the most common metal in the Earth's crust, comes from an **ore** called **bauxite**.

Aluminium has some very useful qualities. As well as being lighter than such **metals** as **iron**, **steel** and **lead**, it is very flexible. It is a good **conductor** of **heat** and **electricity**. And it does not **rust**.

Aluminium is used where these qualities are important - for example: to carry **electricity** in overhead cables; for step ladders that have to be carried from place to place; in aeroplane parts and (as a thin foil) for wrapping food.

Amplitude

When a pendulum is set going, or a guitar string is plucked, it **vibrates**, that is it moves to and fro in a regular rhythm.
The distance the pendulum bob or the string moves from its resting (central) point to one side is called the **amplitude** of the vibration.

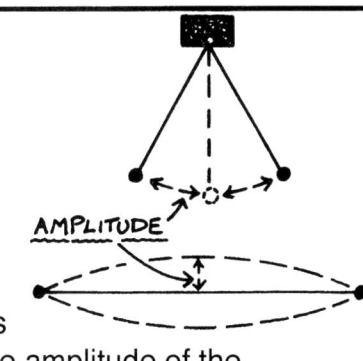

When a guitar string is plucked, or the end of a ruler is twanged, the vibration makes a **sound**. The bigger the amplitude of the vibration is, the louder the sound is.

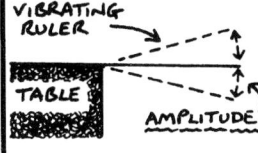

LARGER AMPLITUDE, LOUDER SOUND

SMALLER AMPLITUDE, QUIETER SOUND.

Waves on the sea also move up and down in a regular rhythm. The bigger the amplitude of a wave, the more powerful the wave is

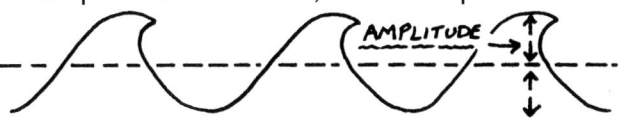

Animal

Animals live almost everywhere on the Earth. Fish in water are animals. Birds and insects in the air are animals. Worms in the soil are animals. Pets are animals, but so are the humans that keep them.

Animals **feed**. They take in food and break it down to get **energy**.
Animals **breathe**. They take in **oxygen** to help them break food down and get energy.
Animals get rid of **waste** material (**liquids** and **solids**).
Animals **move** (the whole animal and parts of the animal).
Animals are **sensitive**. They can see, hear and feel things around them.
Animals **grow** (the whole animal and parts of the animal).
Animals produce new animals - they **reproduce**.

Animals are one group of **living things**. **Plants** are another group of living things.

Antibiotic

People who are ill are sometimes given antibiotics to help them get better.
Antibiotics are made by certain bacteria. The kinds of antibiotic that are used to cure illness work by killing other bacteria - the bacteria that cause the illness.

There are many different antibiotics, and the doctors use different ones to cure different illnesses. One famous antibiotic is penicillin, which was discovered by Sir Alexander Fleming in 1928.

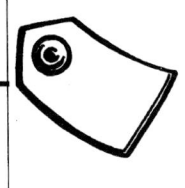

Artery

An **artery** is a tube or **blood vessel** that carries **blood** away from the **heart** to the **tissues** in one of the parts of the body. Examples of arteries are: the **pulmonary artery**, which carries blood to the **lungs**; the **renal artery**, which carries blood to the **kidneys**; and the **femoral arteries**, which carry blood to the **legs**.
The blood goes back to the heart in other blood vessels called **veins**.

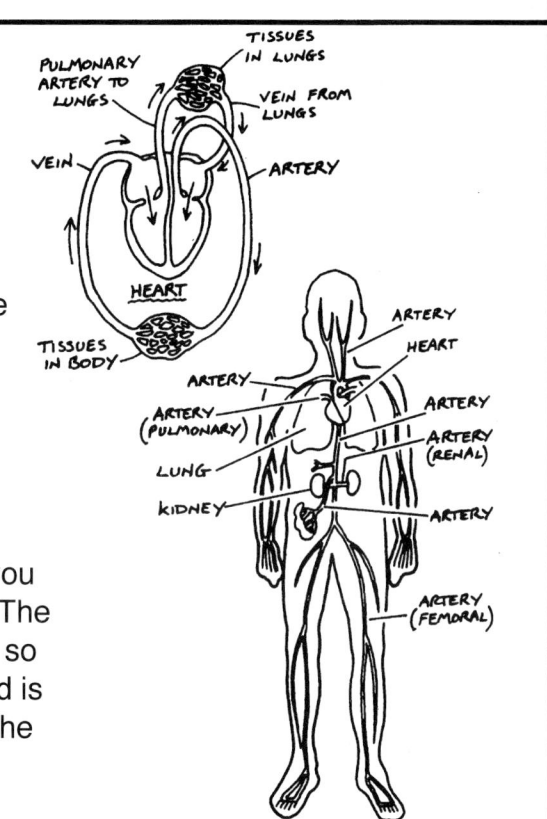

The heart pumps the blood out in spurts (you can feel the spurts if you feel your **pulse**). The arteries are made of thick, elastic **muscle**, so that they can expand when a spurt of blood is coming through and contract again when the spurt has passed.

Atmosphere

The air that all living things **breathe** is part of the Earth's **atmosphere**.

The atmosphere is a **mixture** of **gases** which completely surrounds the Earth. It stretches for many kilometres above the Earth's surface, and eventually thins out into 'empty' space.

The air we breathe, at the lowest level of the Earth's atmosphere, is mostly made up of the gases **nitrogen** (78%) and **oxygen** (20%). There is also a tiny amount of **carbon dioxide** (less than 1%) which is very important for life on Earth. This air also normally contains **water vapour**.

Some moons, some suns and some other planets (for example, Venus and Mars) are surrounded by an atmosphere, but their atmospheres are not like ours.

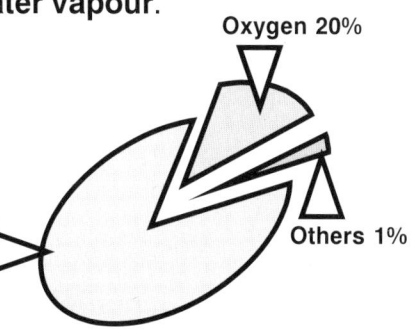

Atmosphere (continued)

The Earth's atmoshere has several different layers, each with a different mixture of gases. Starting from the Earth, the layers are called: the troposphere, the stratosphere, the mesosphere and the thermosphere.

All our **weather** happens in the troposphere.

The **ozone layer**, which stops harmful ultraviolet light rays getting through to the Earth from the Sun, is in the stratosphere.

The **temperature** is different at different heights in the Earth's atmosphere.

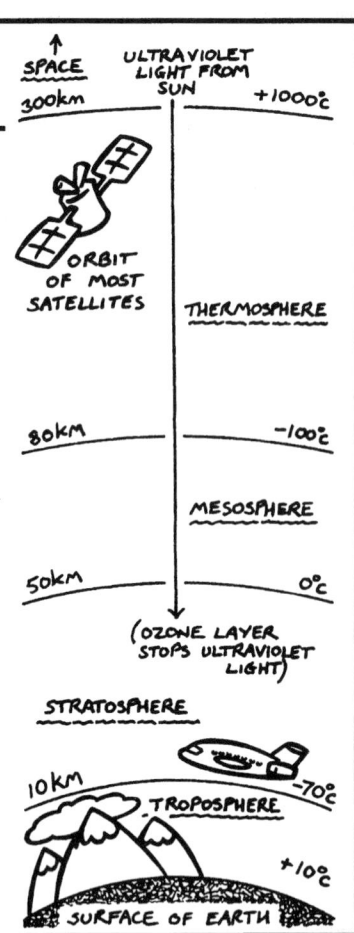

Atom

Two thousand years ago, the ancient Greeks suggested that all **substances** are made up of tiny particles called **atoms**, and we still believe this.

Examples of atoms:	**Symbols for these atoms:**
hydrogen atoms	H
oxygen atoms	O
nitrogen atoms	N
iron atoms	Fe

Atoms are often joined together to make **molecules**. In a water molecule, 2 atoms of hydrogen are joined to 1 of oxygen. So water has the symbol H_2O.
In a salt molecule, an atom of sodium (Na) is joined to an atom of chlorine (Cl).
In the molecules of the air we breathe out, an atom of carbon (C) is joined to 2 atoms of oxygen (O).

We now believe that an atom is made up of a **nucleus** with **electrons** orbiting round it.

Barometer

The air around us exerts a **pressure**. This means that it pushes on everything in the Earth's **atmosphere**.
We don't really notice this air pressure, but it can be measured with a **barometer**.

One type of barometer is made from a long glass tube, closed at one end. The open end is dipped into a container of **mercury**.
The air pressure pushes down on the mercury in the container, and this forces up the mercury inside the tube. The higher the air pressure, the higher the mercury rises up the tube.
When the air pressure is normal, the mercury should rise about 76cm up the tube.

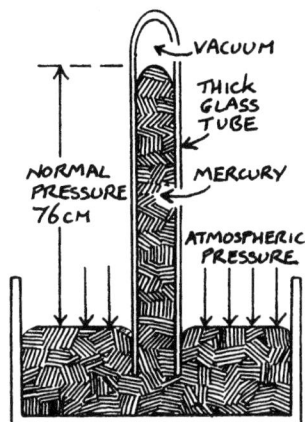

Mercury is used in barometers because it is a 'heavy metal'. If water was used, the column of water in the tube would be about 10 metres high.

Barometer Continued

The first mercury barometer was made by an Italian named Torricelli in 1645.

Nowadays, most barometers are made from a metal box with most of the air inside taken out (leaving a partial **vacuum**). The air pressure outside pushes on the sides of the box.
If the air pressure rises, the box is squeezed more and a pointer connected to the top of the box moves one way.
If the air pressure goes down, the box is squeezed less and the pointer moves the other way.

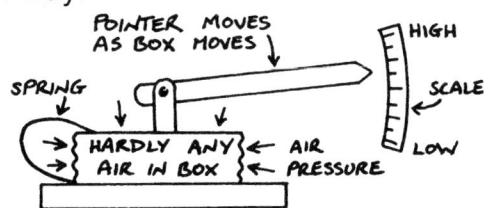

The same type of barometer is used in aeroplanes to measure how high they are flying. As the plane gets higher, the air pressure drops and the needle goes down.
When a barometer is used like this, it is called an **altimeter**.

Biodegradable

Dead leaves break up and become part of the ground. They are **biodegradable**. Apple cores dry out and shrivel up. They are biodegradable. Human bodies are also biodegradable.

If something is biodegradable, it rots away. This rotting process happens because **microbes** such as **bacteria** break down dead **plant** and **animal tissue**.

A garden compost heap is a good example of dead material (for example, left-over cooked vegetables, potato peel and tea leaves) that is being rotted away by microbes. Only biodegradable things should be put on a compost heap.

Things that are **non-biodegradable**, such as most **plastics**, can be difficult to get rid of.

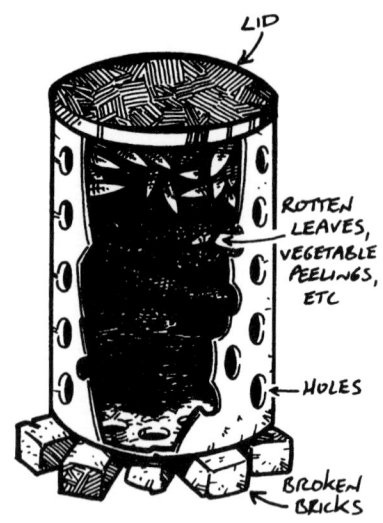

Blood

Imagine nine pints (about 5 litres) of milk on your doorstep. This is roughly how much blood there is in an adult's body.

Blood carries food and oxygen round the body to keep the body cells alive; it helps to take the waste away from the body; and it kills germs.

Blood is made up of red cells, white cells, platelets and plasma.

The red cells carry oxygen away from the lungs to all parts of the body. The colour in them, from a chemical called **haemoglobin**, is what gives blood its red colour.

White blood cells destroy many dangerous germs that enter the body. They do this in 2 ways. Some white cells 'eat' germs.
Other white cells make chemicals called **antibodies** which help to protect the body from disease. Each type of antibody is produced to fight a specific infection.

Blood (continued)

Platelets in the blood help to stop bleeding. When exposed to the air, they clot together and help to form a layer over a cut or a wound. This traps red cells, which 'dry out' to make a solid plug or 'scab'. This stops the bleeding and keeps out dirt and germs.

Plasma is the watery liquid which carries the red and white blood cells and the platelets. It circulates around the body in tubes called **blood vessels**, driven by the beating of the heart.

There are four different **blood groups** or types of blood: called A, B, AB and O. It is vital to know someone's blood group if they are going to be given a **transfusion** of blood from someone else. For example, a person with group O blood can only receive blood which is in group O.

Camera

A **camera** takes pictures by allowing **light** to pass through a small hole at its front and fall onto a piece of film at its back. The film is sensitive to light, and so an **image** of the object in front of the camera is made on the film.

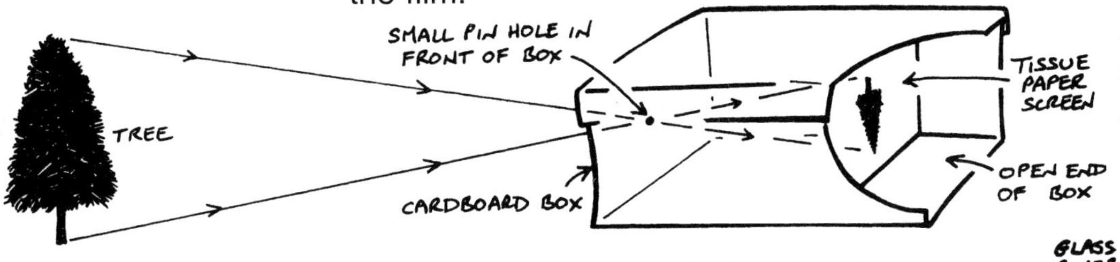

A **pinhole camera** can be made from a cardboard box and tissue paper

The light travels through the pinhole onto the film at the back.
Most cameras have a lens instead of a pinhole. The lens helps to focus the light onto the film.

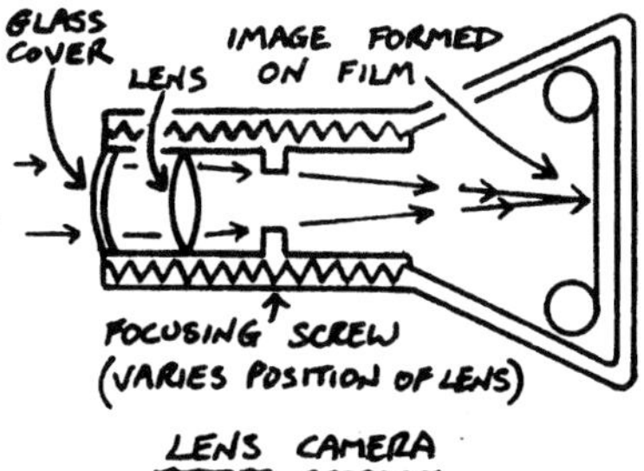

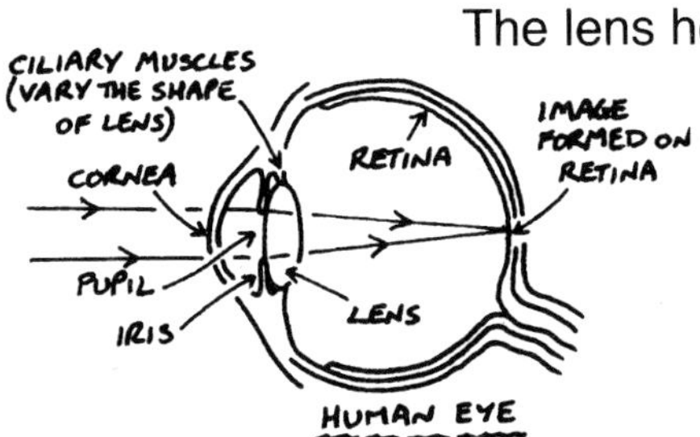

In some ways, the lens camera is like a **human eye.** The eye has a lens at the front and a region at the back which is sensitive to light (the **retina**).

Capillary

A **capillary** is a very narrow tube that carries **blood** through the **tissues** of **animals**.

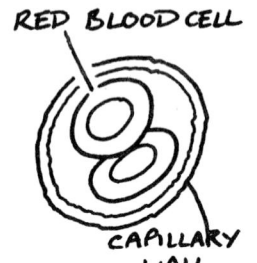

Capillary walls are very thin. They allow fluid to pass through them into and out of the tube.
Oxygen and food dissolved in blood pass out through the walls of the capillary to feed the **cells** of the body.
Carbon dioxide and waste material dissolved in blood pass back into the capillary and are carried away so that the body can get rid of them.

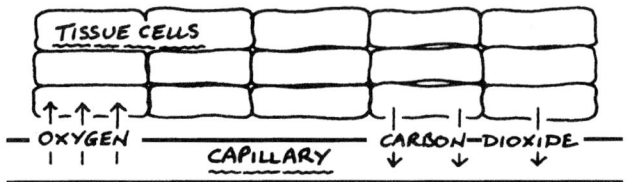

The tiny capillaries join up to form small **veins**, which then carry blood to larger veins.

Carbohydrate

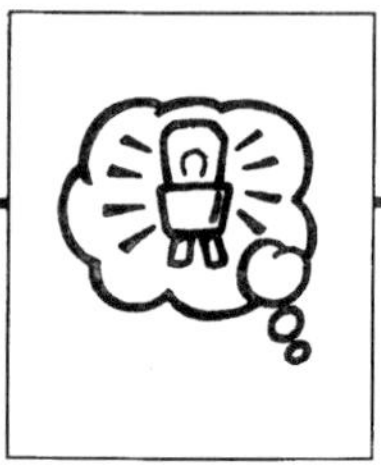

Carbohydrates are food substances made by plants. They contain carbo (**carbon**) hydr (**hydrogen**) ate (which means **oxygen**).

Sugar and **starch** are the main carbohydrates.

Potatoes and grains both contain large amounts of starch, and foods made from grain products, such as bread and pasta, contain a lot of starch.

Sugar cane and sugar beet contain large amounts of sugar, and it is these plants that produce the sugar we add to foods and drinks to make them sweet. Many fruits also contain large amounts of sugar.

We use carbohydrates to give us **energy** – but if we have too much carbohydrate and too little exercise, the carbohydrate we don't need is stored by our bodies as **fat**.

Carbon

What do soot, charcoal and diamonds have in common?
The answer is that they are all forms of **carbon**.

Carbon is an **element** that is found in two pure forms: one black and slippery, called 'graphite'; the other very hard and transparent, called 'diamond'.

Graphite is used in the 'lead' in pencils (pencils don't actually contain lead). Because it is soft and slippery, graphite is also used as a lubricant to help parts of machines slide over each other and overcome **friction**.

Graphite and diamond are so different from each other because in graphite the carbon **atoms** are arranged in layers that can slide easily over each other and in diamond the carbon atoms are arranged in a very strong, crystal-like structure.

Carbon dioxide

Carbon dioxide is one of the **gases** that make up the air we breathe. It contains one **atom** of **carbon**, joined to 2 atoms of **oxygen** (CO_2). The other important gases in the air are oxygen (O_2) and **nitrogen** (N_2). Only a tiny part of the air, much less than one per cent, is carbon dioxide.

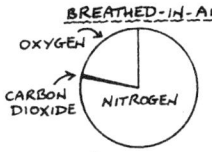 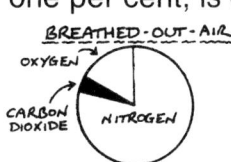

Plants with green leaves need carbon dioxide to live and grow. During the day, they use the carbon dioxide from the air to build up food. As they do this, they produce oxygen which goes back into the air.

Humans and other animals need to breathe in oxygen from the air to stay alive. As we do this, we produce carbon dioxide, which we breathe out.

By converting carbon dioxide back to oxygen, plants help to keep a healthy balance of gases in our atmosphere.

Carbon dioxide (continued)

Most of the fuels we burn (like oil, coal and wood) contain carbon. This carbon burns in air and produces carbon dioxide. We burn far more of these fuels now than we ever did in the past, so there is a danger of upsetting the balance of gases in the air.

Another threat to the atmosphere is the chopping down of trees. Every year, the world loses more trees as land is cleared. This means that there are fewer green leaves to convert carbon dioxide back into oxygen.

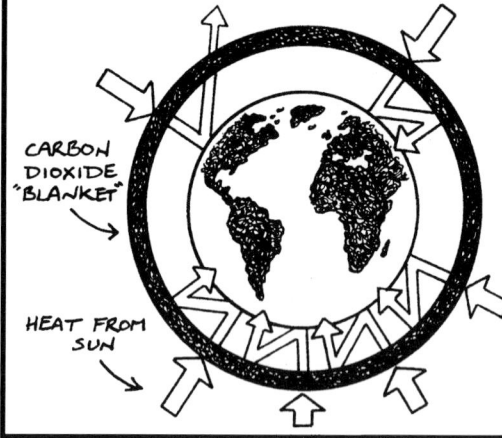

Producing too much carbon dioxide is dangerous, because it acts as a kind of blanket around the Earth. **Radiated heat** from the Sun can get to the Earth through this blanket, but then it gets trapped by the blanket and cannot get away again (the greenhouse effect).
The effect of this is to warm the Earth up, which changes the weather and even melts the ice at the North and South Poles, perhaps causing the sea level to rise.

Carnivore

Animals that eat the flesh of other animals are described as **carnivores**. Some animals eat meat (flesh), some eat only vegetables, and some eat a mixture of meat and vegetables.

Catalyst

When two or more chemical substances are put together in order to get a **chemical reaction**, it may take some time for the reaction to take place. Sometimes another chemical can be added to speed up the reaction, and this extra chemical is called a **catalyst**.

Although the chemical which is the catalyst in a reaction helps to speed up the change between other chemicals, it is not used up itself.

The chemicals in a washing powder react with the dirt on the clothes in a washing machine, and this reaction lifts the dirt out of the clothes.
Some washing powders contain extra chemicals called **enzymes**. These act as catalysts, helping the other washing powder chemicals break up some kinds of dirt.

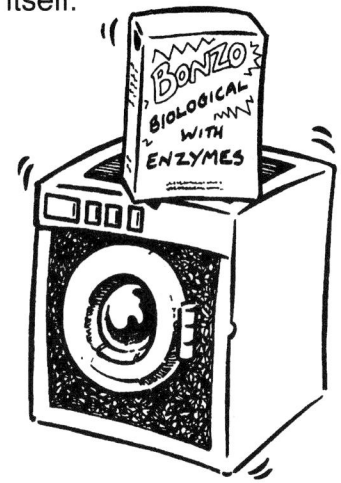

Cell

All living organisms, whether they are **plants** or **animals**, are made up of small **cells** that can only be seen with a **microscope**.
There are some living organisms that consist of just one cell, but most organisms consist of millions and millions.

Organisms are made up of different **tissues**. The cells in each tissue are different from those in the other tissues and do different jobs. Examples of animal body cells are: red blood cells which carry oxygen round the body; fat cells, which store fat; and nerve cells, which carry messages round the body.

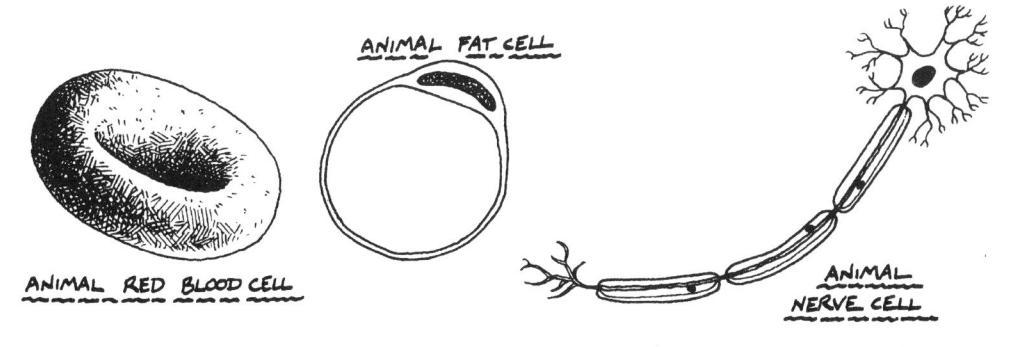

ANIMAL RED BLOOD CELL

ANIMAL FAT CELL

ANIMAL NERVE CELL

Cell (continued)

All cells consist of:
a **membrane** round the outside, through which substances pass in and out;
a clear, jelly-like substance called **cytoplasm** inside the membrane;
and a **nucleus** inside the cytoplasm.

Plant cells have a membrane, cytoplasm and a nucleus, but they also have:
chloroplasts – specks which contain **chlorophyll** and give the plant its colour;
a tough **cell wall** made of **cellulose**;
a **vacuole** in the centre which stores **cell sap**. This pushes out towards the cell wall and helps to give the cell and the plant a firm shape.

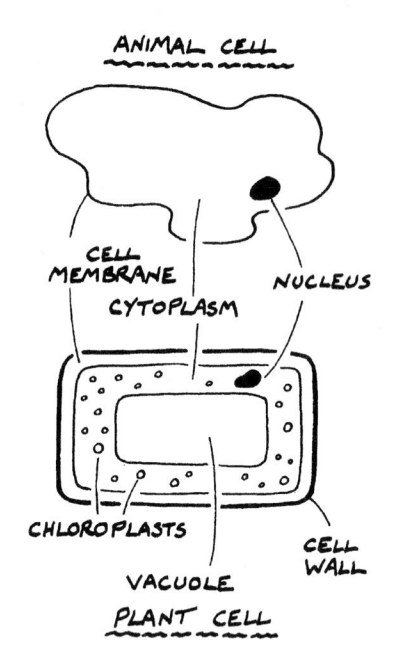

Celsius

The **Celsius scale** is the scale that most people use to measure the temperature of something – to say how hot or cold it is.

The Celsius scale is named after the Swedish scientist, Anders Celsius, who decided it would be helpful to have a scale of 100 equal parts (**degrees**) between two fixed temperature points: the lower point, at which water freezes, (0°C) and the higher one, at which water boils (100°C).

Because the distance between the two fixed points on the scale has been divided into 100 degrees, this scale has, in the past, been called the centigrade scale.

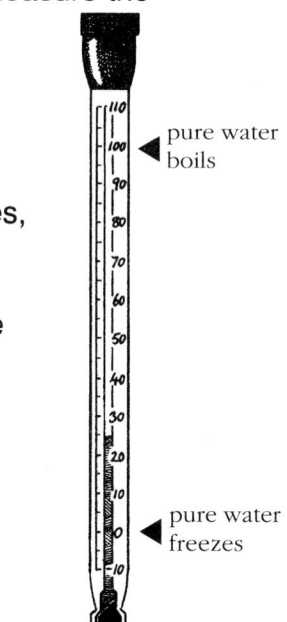

Chemical reaction

When two or more chemical substances are mixed together, there is often a **chemical reaction** between them. This means that the **atoms** that were joined together in the original substances break apart and rearrange themselves to make a new substance. This new substance is quite different from the original substances.

Some chemical reactions produce **heat**, some produce **light**, some produce **sound** and some produce changes in **colour**.

A sparkler contains **magnesium**. This, when lit, reacts with **oxygen** in the air and produces light and heat.

The chemicals inside a banger or a rocket, when lit, react with the oxygen in the air and produce heat, light and sound.

Chlorine

Chlorine is a **gas** with a light yellow-green colour. It is poisonous and attacks the **tissue** in the **throat** and the **lungs** if people breathe it in.

Very small amounts of chlorine are added to **water** at water treatment works. This chlorine acts as a disinfectant and kills **bacteria**, making the water safe to drink.

Chlorine is also added to the water in swimming pools to make it safe, but it may make your eyes sting a little when you swim in it.

Chlorophyll

Chlorophyll is a green substance that is contained in the **chloroplasts** of **plant cells**. It gives plants their green colour.

The chlorophyll uses the **energy** in sunlight to turn **carbon dioxide** from the air, with **water**, into **oxygen** and **carbohydrates** (sugar and starch).

This process is called **photosynthesis** from the words 'photo', which means 'using light' and 'synthesis' which means 'building up'.

It is the carbohydrates produced by photosynthesis that form the basis of all the food of all the living things in the world. People cannot make food from air and water; plants can and do.

Circuit

A **circuit** is any path or track which finishes up where it started from. For example, in motor racing, cars go round and round a circuit. The path of a circuit can be long or short, straight or winding, but it must finish where it started.

An **electric circuit** is the path that an **electric current** travels around. If the path is not a complete circuit, no electric current will flow through it. A **switch** can be used to make a gap in an electric circuit and stop the flow of electricity.

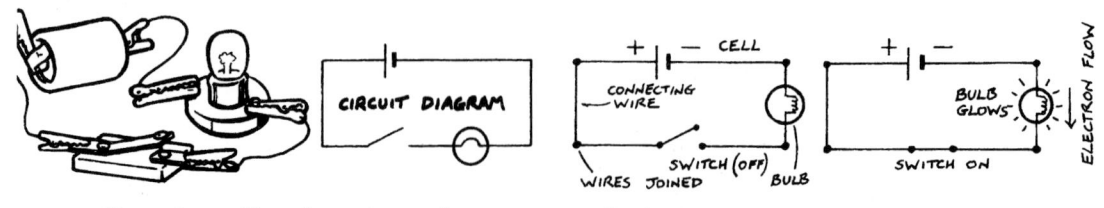

Drawings like the ones above are called **circuit diagrams**.
Each part of the circuit has its own special symbol, e.g. there is a symbol for a light bulb, another for an **electric cell** or **battery**, another for a switch, etc.

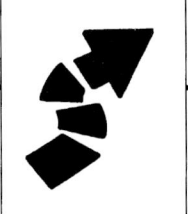

Combustion

Combustion is another word for burning. It is one of the most important processes that happens on Earth.

Combustion is a chemical reaction between a gas (usually **oxygen** from the air) and the chemicals in a **fuel** - for example, the **hydrogen** and **carbon** in **fossil fuels** (coal, natural gas and oil).

When fossil fuels burn, they produce **carbon dioxide**, **water** and **heat**.

FOSSIL FUEL + OXYGEN = CARBON DIOXIDE + WATER → HEAT

We put the heat from combustion to many uses, for example for gas fires and cookers at home, for generating electricity in power stations, and for making car engines go (burning petrol).

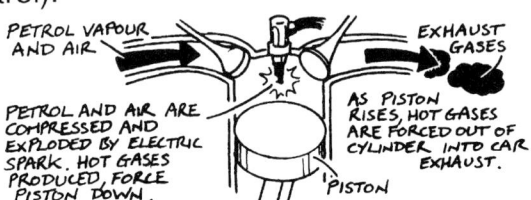

PETROL VAPOUR AND AIR

PETROL AND AIR ARE COMPRESSED AND EXPLODED BY ELECTRIC SPARK. HOT GASES PRODUCED, FORCE PISTON DOWN.

EXHAUST GASES

AS PISTON RISES, HOT GASES ARE FORCED OUT OF CYLINDER INTO CAR EXHAUST.

PISTON

Combustion (continued)

Combustion gives us problems as well as advantages in our everyday lives.

Carbon dioxide is produced by the burning of the carbon in fossil fuels, and the huge amount of this gas that we are now sending into the atmosphere may be adding to **global warming**.

INCLUDES CO_2 (CARBON DIOXIDE)

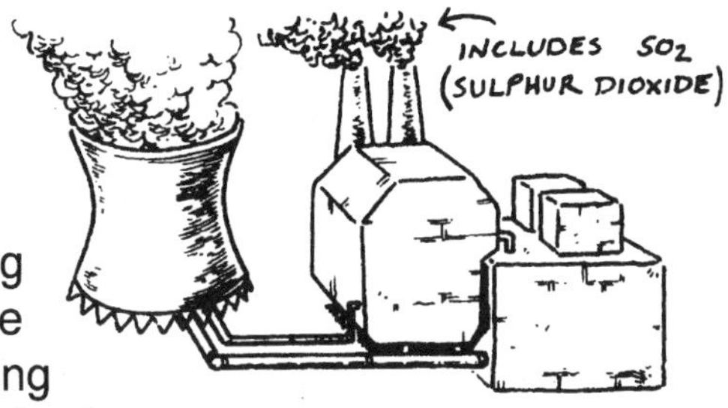

INCLUDES SO_2 (SULPHUR DIOXIDE)

Sulphur dioxide is produced by the burning of the **sulphur** in fuels such as coal, and the large amount of this gas that we are releasing returns to earth from the atmosphere in **acid rain**.

Very rapid combustion causes an explosion, for example in fireworks.

Compounds

When two or more **elements** join together chemically, they make a new substance called a **compound**.

A compound is totally different from the elements that it comes from. For example, **carbohydrates** are quite different from each of the elements they contain - **carbon**, **hydrogen** and **oxygen**. It is not possible to separate these elements, but if you burn a piece of toast (bread contains carbohydrates), you can see the black carbon in it.

A compound is different from a **mixture**. If two or more substances are mixed, they do not become chemically joined, and each keeps its own properties. For example, if **iron filings** and **sulphur** are mixed, they can be separated again with a magnet. (The iron sticks to the magnet, but the sulphur does not.)

Hydrogen and **oxygen** can be combined to form a compound, H_2O, which is water.

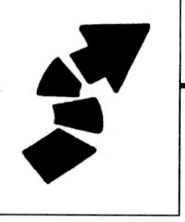

Compression

If you press something together or squeeze it so that it takes up a smaller space (has a smaller volume), you **compress** it.

You can compress a sponge or a pillow (foam or feather).

Some things are easier to compress than others.

If you trap some **air** in a syringe or a bicycle pump, you push on the plunger and make the volume of the air much smaller.

But if you fill a syringe with **liquid** (e.g. water) and try to compress it, it hardly changes its volume at all.

Liquids are almost 'incompressible'. This is why they are used in **hydraulic** systems. If you press on the liquid at one part of the system, e.g. a car brake, the **pressure** is passed on through the liquid to another part of the system, e.g. the brake pad next to the car wheel.

Concave

Concave means curving inwards. It has the opposite meaning to **convex**.

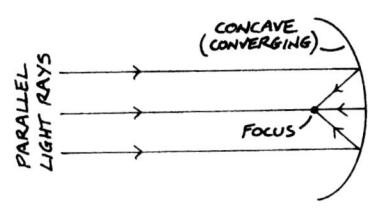

Some **mirrors** are concave. They reflect light inwards towards a **focus** (they **converge** the light rays). If you put an object close to the mirror, its image in the mirror is magnified. For this reason, this type of mirror is used to help people shave or make up their faces.

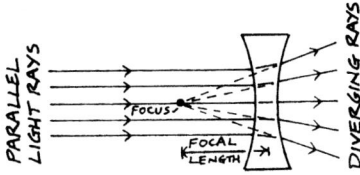

Some **lenses** curve inwards. They are called concave lenses. A concave lens spreads out light rays (it **diverges** them). It does the opposite of what a concave mirror does. Concave lenses are used in spectacles for people who are short sighted, to help focus light rays onto the **retina** at the back of their eyes.

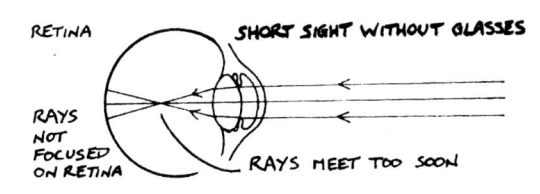

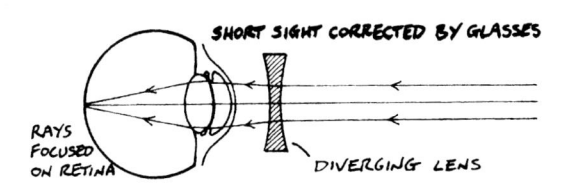

Constellation

The pattern formed by a group of stars as we see them is called a **constellation**. The constellations we look for in the night sky were usually first observed thousands of years ago, but they have changed (very slowly) since then.

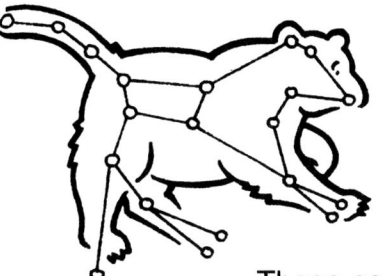

Astronomers in ancient times drew an imaginary line from star to star in a constellation and created a picture which they used to give the constellation a name. For example, when the stars in the Great Bear were joined by a line, the picture made the astronomers think of a large bear.

These constellations can be very difficult to see. But in the northern hemisphere, most people can spot the Plough (a smaller constellation within the Great Bear) by looking north. (Remember that in ancient times, ploughs were long tools that were pulled along through the soil by farmers.) The Plough is also called the Big Dipper - you can see the ladle shape.

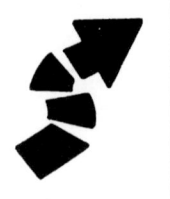

Condensation

When a **gas** or **vapour** changes to a **liquid**, it **condenses**.

We often see condensation happening when **water vapour** from the **air** condenses onto a cold surface. Examples of condensation: little drops of moisture on the inside of a window pane on a cold day when you breathe warm (damp) air on it; and dew on the grass at night when the temperature drops.

If warm air is cooled, it cannot hold as much water vapour as when it was warm, so some of this vapour may turn into water.

Water is not the only substance which is produced as a result of its gas form condensing into its liquid form. Other substances condense, but water is the substance we see condensed most often on this planet.

Conduction

Conduction is the name given to the transfer of **heat** or the movement of **electricity** through a material.

Materials that are good conductors of heat have their **atoms** arranged in such a way that heat is transferred easily from one particle to another. Many metals are good conductors of heat (**copper** and **silver** are particularly good). Other materials, such as wood and plastic, are not good conductors of heat.

Metal pans often have a wooden or a plastic handle, because a metal handle would conduct the heat from the pan to the cook's hand and burn it.

Metals are also good conductors of electricity. This is because they have **free electrons** in their atomic structure. The metal wires used in electricity cables have to be covered in a material such as plastic that does not conduct electricity, to prevent people touching the wire and getting an electric shock.

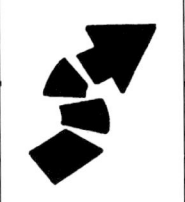

Convection

Convection is the name given to the movement of particles in a **liquid** or **gas** which transfers **heat**.

When a liquid or gas is warmed, it expands. This makes it less dense. The colder, denser liquid or gas sinks, and forces the warmer material upwards, forming a **convection current**. The upward movement of the warmer liquid or gas carries heat with it.

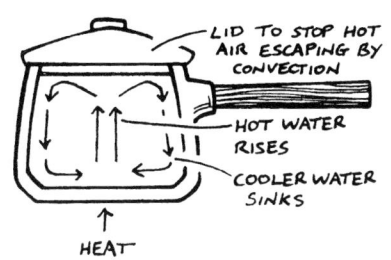

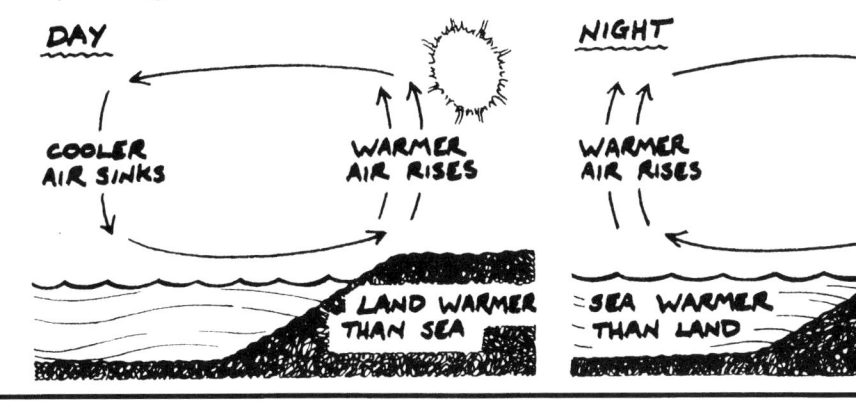

Convex

Convex means curving outwards ('bulging'). It has the opposite meaning to **concave**.

Some **mirrors** are convex. When light reaches them in a beam of parallel **rays**, they spread the rays out (they **diverge** them). If you look into a convex mirror, the things you see will look smaller than they really are, but you will be able to see over a wide angle.

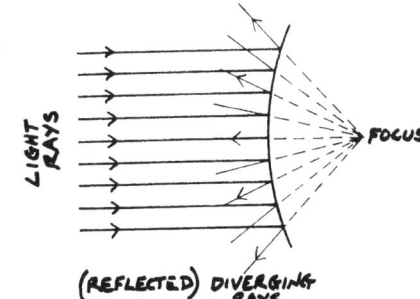

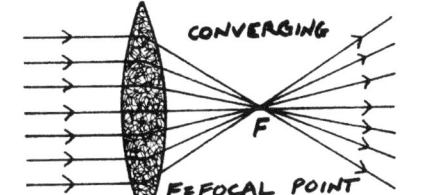

For this reason, this type of mirror is used as a driving mirror.

Some **lenses** curve outwards. They are called convex lenses. A convex lens focuses light rays to a point (it **converges** them), so it does the opposite of what a convex mirror does.

Convex lenses are used as magnifying glasses. They are also used in spectacles for people who are long-sighted, to focus light rays onto the **retina** at the back of their eyes.

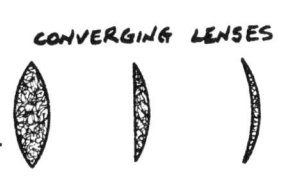

Crystal

Crystals are **solids** with definite, regular shapes. They can be **elements** or **compounds**.
Examples of crystals are common salt and diamonds. Quartz is also a substance that has a **crystalline** form.

The edges of crystals are straight, and the surfaces are flat. Crystals come in many different shapes: for example, they may be cubic or hexagonal.

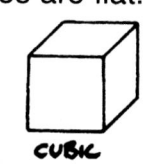

CUBIC   HEXAGONAL

The process of forming crystals is called **crystallisation**.

Some crystals are formed when a **solution** of the chemical dries up. Quick drying usually gives very small crystals and slow drying gives larger ones. This often shows well with concentrated salt solution.

Current

Static electricity can be made by rubbing certain things together, e.g. by rubbing a plastic comb or a balloon against your sleeve. The static electricity comes from **electric charges** (+ or -) that are standing still ('static'). They are standing still because they are on materials that are **non-conductors.**

When electric charges move, they make an **electric current**. An electric current always needs a complete conducting pathway to travel along. This pathway is called a **circuit**.

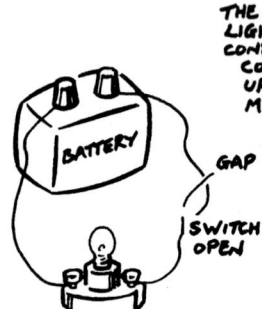

THE BULB ONLY LIGHTS UP IF A CONDUCTOR IS CONNECTED UP BY METAL CLIPS

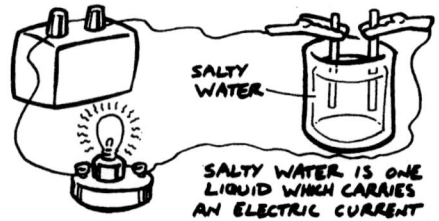

CLIPS
COPPER WIRE
BATTERY
GAP
SWITCH OPEN

SALTY WATER
SALTY WATER IS ONE LIQUID WHICH CARRIES AN ELECTRIC CURRENT

An electric current cannot travel through an **insulator** (a non-conductor). An electric current is a movement or flow of **electrons**.

IF A CIRCUIT IS BROKEN, ELECTRONS CANNOT FLOW. THE CURRENT STOPS.

If an electric circuit is broken, the flow of electrons stops.

Decompose

Things **decompose** when they break up or decay. Plants and animals decompose when they die and their bodies rot.

Rotting is caused by living **organisms** such as **bacteria** and **fungi**. These organisms are called **decomposers**. Decomposers feed on dead animals or plants and break them down into simpler materials. These materials go into the soil and are absorbed by plants, helping them to grow.

Decomposers play an important part in keeping soil fertile and plants healthy, providing animals with food. They also do an essential job by removing the dead material which would otherwise pile up on the surface.

Materials are described as **biodegradable** if they can be decomposed (degraded) in this way. For example, apple cores are biodegradable, but metals in general are not.

Some chemical compounds decompose when they are heated. E.g. if small logs are piled up and covered so that they do not have enough air (oxygen) to burn completely, and a small fire is started at the bottom, water escapes from the logs as steam, gases escape as smoke, and carbon is left as charcoal.

Diaphragm

Your **diaphragm** is a 'sheet' of muscle separating your chest and lungs from your abdomen.

Air is pumped in and out of your lungs by your **diaphragm muscle**. As you breathe in, your diaphragm **contracts** and becomes flatter. This allows air down through your windpipe and into your lungs, under pressure from the atmosphere (the air outside).

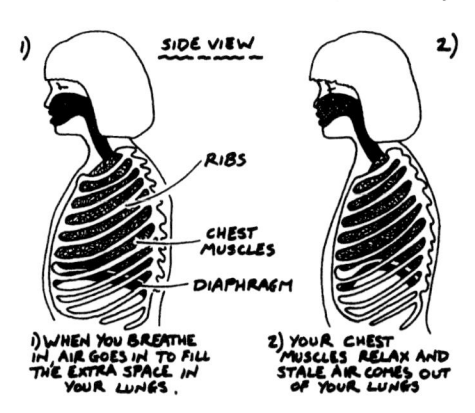

When you breathe out, your diaphragm relaxes into a curved, dome shape. This pushes air upwards and out of your lungs.

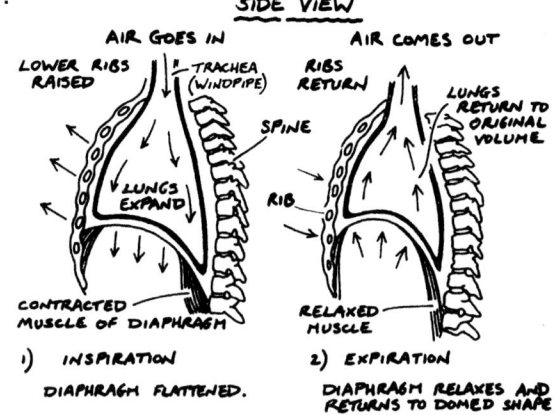

You also have muscles between your ribs which lift the lower ribs when you breathe in. They relax as you breathe out.

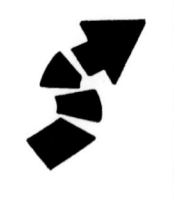

Diffusion

If a gas escapes into the air, it gradually spreads out into the air and all the molecules get mixed up; this spreading out is called **diffusion**.

The exhaust fumes from a car diffuse into the air.

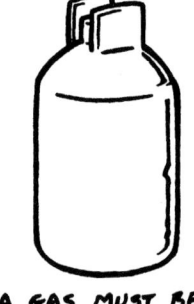

A gas has to be kept in a container, otherwise it diffuses into the air. Light gases, such as **helium** and **hydrogen**, diffuse very quickly. A helium-filled balloon goes soft quicker than an air-filled balloon because the helium gas diffuses out through the walls of the balloon more quickly.

Diffusion also happens in **liquids**. A drop of **ink** left in a beaker of water gradually diffuses through the water.

Diffusion also happens when **oxygen** and **carbon dioxide** move into and out of the cells of the body.

A GAS MUST BE KEPT IN A CONTAINER, OR IT WILL DIFFUSE INTO THE AIR.

Digestion

Pieces of potato, cabbage or bread are far too big and solid to go into your blood and enter the cells of your body. Food needs to be made into simple, soluble substances so that it can pass into your blood, then from your blood into your cells. **Digestion** is the process that does this.

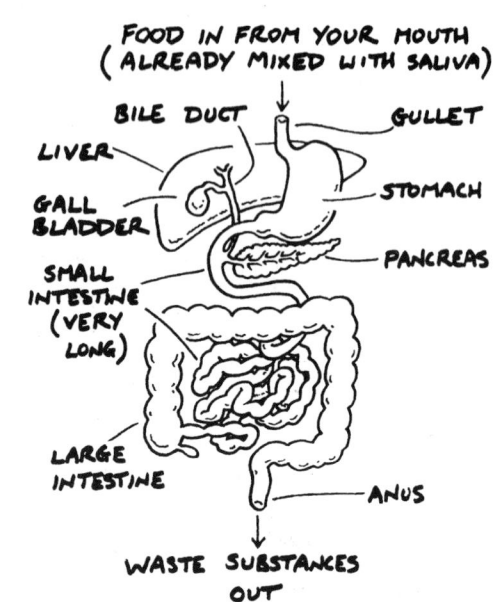

Digestion starts in the mouth. As you chew your food, it mixes with **saliva**, which begins to digest starch. Your tongue pushes food to the back of your mouth, towards the **gullet**, which pushes food on, down to your **stomach**. The food is held in the stomach for about an hour and mixed with **digestive juices**.

The food then goes to the **small intestine** and mixes with **bile** (from the **liver**) and **enzymes** which digest starch, sugars, proteins, fats and oils.

The dissolved food substances go through the gut wall into blood vessels which take them to the rest of the body.

In the **large intestine**, much water is reabsorbed, and then waste or undigested food is pushed out of your body, through the **anus**.

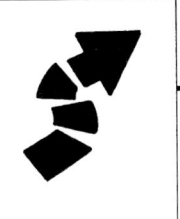

Dissolve

Many things **dissolve** in water - **salt** and **sugar** are the two most common substances. If you drop a sugar lump or a spoonful of salt in water it seems to disappear, leaving a clear liquid. You can still taste the sugar or salt though. Its molecules have moved into the spaces between the water molecules.

The sugar and water molecules mix together to make a **solution**. The water is called the **solvent** and the sugar is called the **solute**:

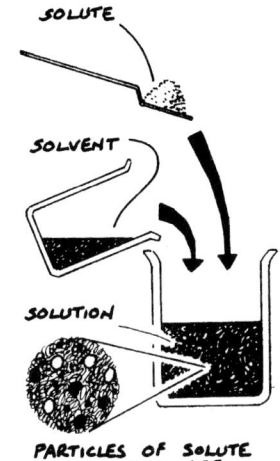

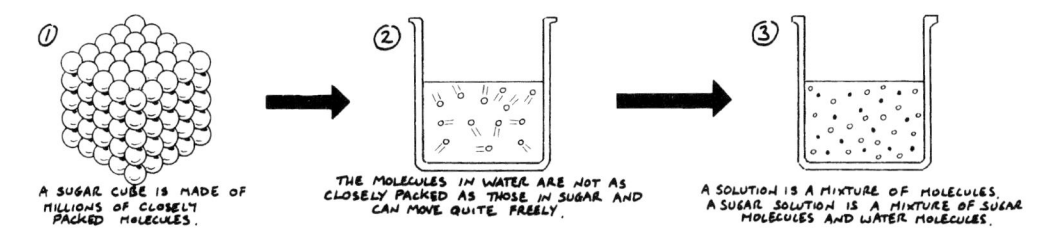

① A SUGAR CUBE IS MADE OF MILLIONS OF CLOSELY PACKED MOLECULES.

② THE MOLECULES IN WATER ARE NOT AS CLOSELY PACKED AS THOSE IN SUGAR AND CAN MOVE QUITE FREELY.

③ A SOLUTION IS A MIXTURE OF MOLECULES. A SUGAR SOLUTION IS A MIXTURE OF SUGAR MOLECULES AND WATER MOLECULES.

SOLUTE
SOLVENT
SOLUTION
PARTICLES OF SOLUTE AND SOLVENT ARE EVENLY MIXED.

If a solution has a lot of solute in it, we say it is concentrated. If it has very little solute dissolved in it, it is **dilute**.

Water is our most common solvent, but there are many others. Not everything dissolves in water, and so a solvent like 'white spirit' or turpentine can be used.

Echo

Sounds can be **reflected** when they meet a solid barrier, like a wall or cliff. An **echo** is a reflected sound wave.

Although echoes can be useful they can sometimes be a nuisance in a room or a hall. Concert halls, theatres and cinemas often have specially padded walls and ceilings. The soft padding absorbs the sound and stops sound waves from echoing round the room, interfering with each other. The study of echoes in halls and theatres is called **acoustics**.

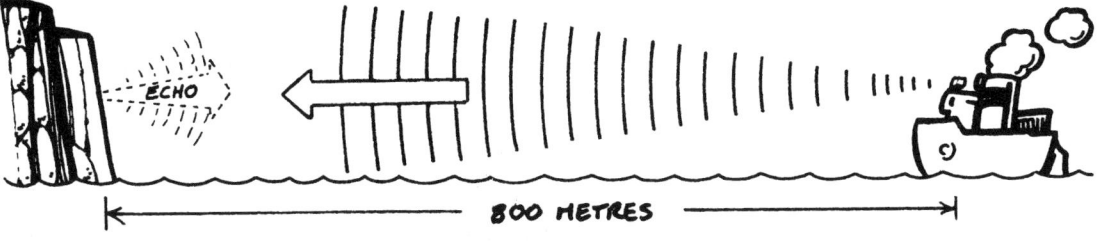

800 METRES

Echoes can be used to measure the speed of sound. Suppose a ship sounds its siren when it is 800 metres from a cliff and an echo is heard five seconds later. The sound travels 1,600 metres, to the cliff and back, in five seconds.

Speed of sound = Distance/Time = 1,600 m/5s = 320 m/s

Ecosystem

Plants and animals (organisms) live together in a **community**. The community, the habitat they live in and their environment are together called an **ecosystem**. The water, plants, mud, rocks and animals in a pond make up an ecosystem. *All* the organisms depend on each other and their habitat to stay alive, and they are linked by food chains and food webs.

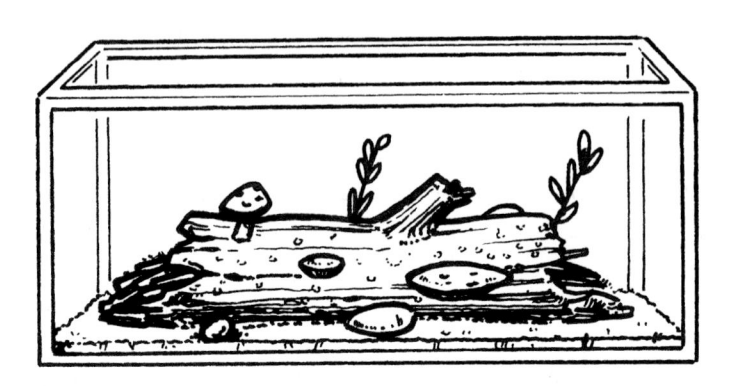

An ecosystem is a self-contained unit, in which all the various organisms live together and interact with each other. A pond, a stream, a forest, a wood or even a rotting log within it can be called an ecosystem.

Some people refer to the Earth as an ecosystem. The study of ecosystems is called **ecology**.

Electromagnet

If you wrap a coil of wire around a nail and pass an electric current through the wire, it becomes an **electromagnet**. The nail will attract small iron and steel objects and has a magnetic field around it like any other magnet. The bigger the current, the stronger the magnetic field; the more the turns of wire, the stronger the magnet.

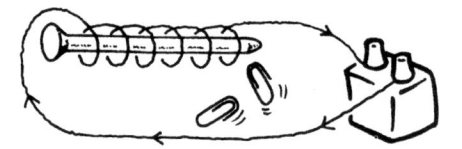

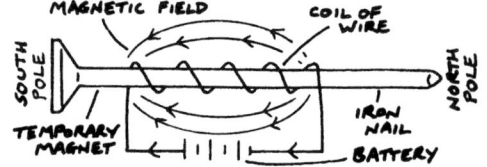

Electromagnets are **temporary** magnets; when the current stops they are no longer magnets.

Electromagnets come in all shapes and sizes and can be very useful: in a telephone, an electric bell, or for lifting heavy objects in scrapyards and dropping them again by switching the current off.

Electron

Everything is made up of tiny particles called **atoms**. The centre of an atom is called the **nucleus**. Around the nucleus are even smaller particles called **electrons**. These travel round the nucleus in a kind of orbit. They exist in layers or shells.

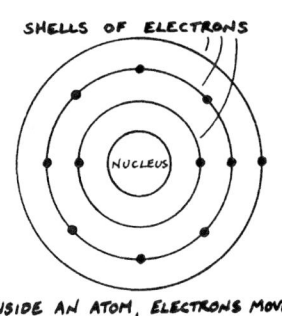

The shells are not really there - they are just the paths taken by the electrons as they move around the nucleus.

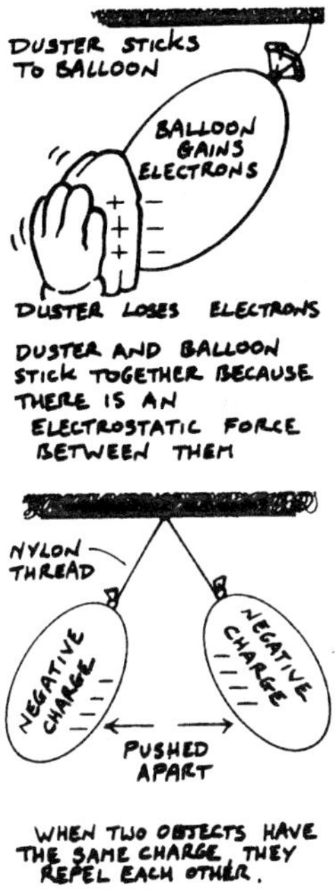

The force holding the nucleus together is very strong, but the force holding the electrons onto an atom is much weaker. Electrons can be taken away from some atoms just by rubbing. When you rub a balloon with a duster the duster sticks to the balloon. Rubbing removes electrons from atoms in the duster and gives them to the balloon. All electrons carry a negative charge. The balloon becomes negatively charged with its extra electrons, and the duster becomes positively charged. The positive and negative attract each other. This is **static electricity** - the electrons are not moving.
(Moving electrons form an electric current.)

SHELLS OF ELECTRONS
NUCLEUS
INSIDE AN ATOM, ELECTRONS MOVE AROUND A CENTRAL NUCLEUS.

DUSTER STICKS TO BALLOON
BALLOON GAINS ELECTRONS
DUSTER LOSES ELECTRONS
DUSTER AND BALLOON STICK TOGETHER BECAUSE THERE IS AN ELECTROSTATIC FORCE BETWEEN THEM

NYLON THREAD
NEGATIVE CHARGE
NEGATIVE CHARGE
PUSHED APART
WHEN TWO OBJECTS HAVE THE SAME CHARGE, THEY REPEL EACH OTHER.

Element

All substances are made up of **atoms**. If you could see inside a gold ring, you would find that all the atoms are exactly the same - they are all identical gold atoms. A substance which has atoms that are all the same is called an **element** (substances made out of different atoms joined together are called **compounds**.).

Another way of describing an element is that it is a pure substance which cannot be split into any simpler substances by a chemical reaction (unlike elements, compounds *can* be split into simpler substances).

There are just over 100 elements that we know of. Each element has its own shorthand symbol for one atom.

Although there are over 100 elements, only five of them make up most of the Earth's crust.

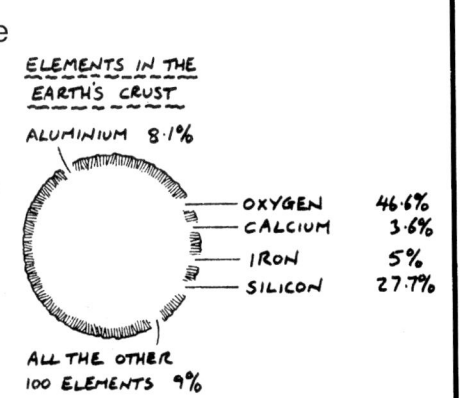

SOME COMMON ELEMENTS AND THEIR SYMBOLS

HYDROGEN	H	SODIUM	Na	IRON	Fe
CARBON	C	SULPHUR	S	GOLD	Au
NITROGEN	N	CHLORINE	Cl	LEAD	Pb
OXYGEN	O	POTASSIUM	K		

ELEMENTS IN THE EARTH'S CRUST
ALUMINIUM 8.1%
OXYGEN 46.6%
CALCIUM 3.6%
IRON 5%
SILICON 27.7%
ALL THE OTHER 100 ELEMENTS 9%

Embryo

An **embryo** is a new developing individual (plant or animal).

The embryos of mammals develop inside the body of the mother. A human embryo develops in its mother's womb for a total time of about nine months after **fertilisation** by a sperm.

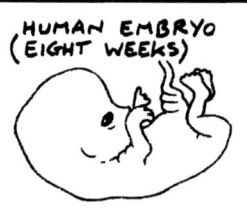

With some animals, such as snakes and birds, the embryo develops inside an egg which has been laid by the mother. After fertilisation, while the egg is still inside the mother bird, a shell develops around it. The shell protects the embryo and stops it from drying up after the egg has been laid. Fertilised eggs hatch – eggs for eating have not usually been fertilised.

With some animals (frogs and many fish) the eggs at first contain just an **ovum** (a female sex cell) in each and the male deposits **sperm** (male sex cells) over them after they have been laid.
Then the sperms and eggs join up, one to one.

A plant embryo is contained inside a seed - each seed has a hard wall around it to protect the embryo and a store of food inside it.

Energy

It is hard to say exactly what **energy** is. You often say that somebody is 'full of energy'. People need energy for walking, running and working - their energy comes from the food they eat. But many other things have energy, too: the wind, running water, a coiled spring, gas and oil, machines, the Sun . . . Energy is what makes things happen.

In science, energy is the ability to get things done, to make things go or move, or the capacity to do work. It is measured in **joules**.

It is often useful to divide energy up into different forms. Two forms of energy that come directly from the Sun are **heat** and **light energy**. **Potential energy** is the name of energy which is stored up waiting to be used. **Chemical energy** is one type of stored energy in fuels and food. If you bang a drum, speak or play a guitar, you are setting free another type of energy: **sound energy**. When something is moving, like a runner, a car or a train, this gives it a form of energy known as **kinetic energy** (movement energy).

Energy (continued)

Electrical energy is changed into light energy by a light bulb, and into heat energy by an electric heater.

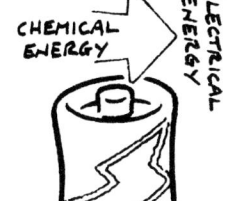

Chemical energy is changed into electrical energy in a torch battery.

Kinetic energy is changed into heat energy when you rub your hand on your sleeve.

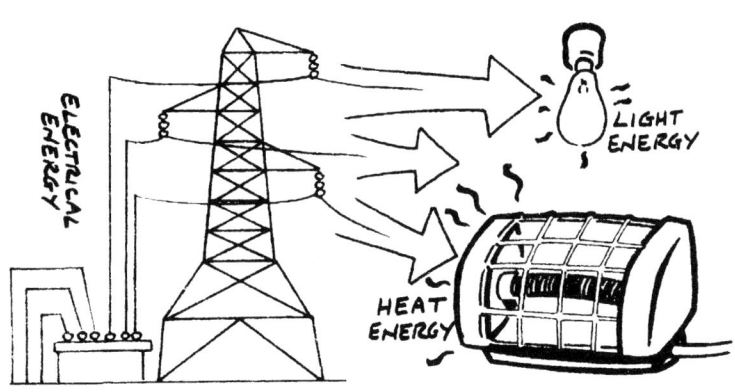

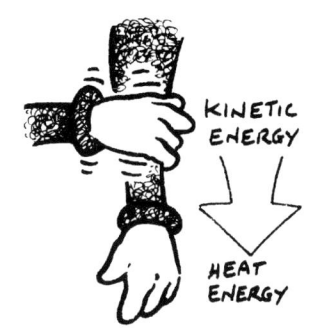

Enzyme

Enzymes are special proteins, found in all living things, which are vital to the chemical reactions of life. Enzymes act as **catalysts** - this means that they speed up chemical reactions without being changed themselves. There are lots of types of enzymes, for example:

Digestive enzymes help to break down our food into simpler substances. Our saliva, and the juices in our stomach contain enzymes.

Respiratory enzymes inside cells help these simpler substances to release energy in the cells of our body. Other enzymes in the cells build up new proteins.

Some detergents contain enzymes which help to remove stains, like sweat, milk, chocolate and blood, from your clothes. These are often called 'biological detergents'.

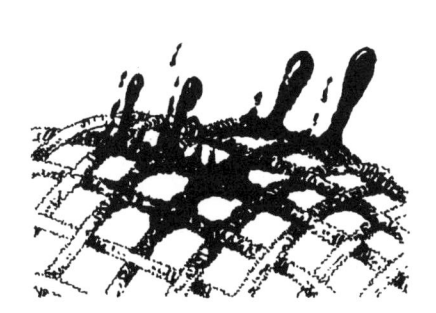

Equilibrium

If something is in balance it is in **equilibrium**. This tight rope walker is in equilibrium - the long pole is helping him to balance.

rope

Some objects are hard to topple over. The stand on the left has a very heavy base, so that if you tilt it to one side it comes back upright again. It is very **stable**. A Russian doll can be made so that it is very difficult to topple over.

Racing cars are built so that their wheels are far apart and the weight of the car is very low down - they have a low 'centre of gravity'. Racing cars are very stable.

Evaporation

When a puddle of water dries up on a warm day, it **evaporates**. The water in the puddle was not boiling but it did gradually change from a liquid to a vapour. This is called **evaporation**. The water vapour goes into the air.

Sometimes, the water vapour held in the air changes back to a liquid. You see this on the inside of a house or car window when it is very cold outside. The water vapour has **condensed**. **Condensation** is the opposite of evaporation.

Water is evaporating and condensing all the time on Earth in a cycle called the **water cycle**. Water from a lake or the sea and from fields and trees evaporates, then rises and cools down. The water vapour condenses to form a cloud and water often falls from the cloud as rain:

Evolution

The theory of evolution states that the living things we see on Earth today did not just 'suddenly appear'. They were produced over billions of years by very gradual changes from organisms which have lived in the past. All life on Earth has **evolved** from life in the past.

EQUUS (THE MODERN HORSE) — RIGHT FORELEG

MESOHIPPUS LIVED ABOUT 30 MILLION YEARS AGO

The theory was suggested by **Charles Darwin** (1809-1882). Most of the evidence for it comes from the study of **fossils**. For example, fossils of the bones of horses showed that they gradually lost toes and developed a hoof. We can picture how horses have evolved over the last 60 million years.

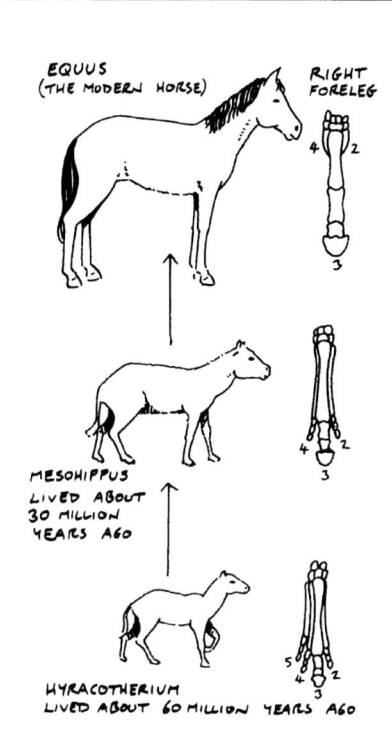

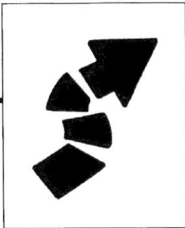

HYRACOTHERIUM LIVED ABOUT 60 MILLION YEARS AGO

Evolution (continued)

Other fossil records have been found for animals like elephants and giraffes. Sometimes these fossil records make a sequence of fossils showing how present-day organisms have evolved from ancestors who lived millions of years ago.

Human beings have also evolved over millions of years from ancestors which looked like apes:

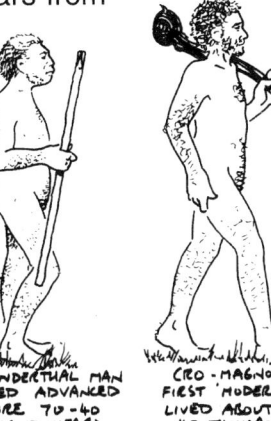

RAMAPITHECUS — EARLIEST KNOWN APE-MAN LIVED 12-14 MILLION YEARS AGO.

RUDOLF MAN — PROBABLY FIRST TRUE HUMAN. A TOOL-USER, LIVED 2½ MILLION YEARS AGO

HOMO-ERECTUS — HUNTER AND FIRE-MAKER, LIVED ABOUT 1 MILLION YEARS AGO

NEANDERTHAL MAN — FORMED ADVANCED CULTURE 70-40 THOUSAND YEARS AGO

CRO-MAGNON MAN — FIRST 'MODERN' MAN. LIVED ABOUT 40 THOUSAND YEARS AGO

As organisms have **evolved** over time, some species have died out. Other things have changed or adapted to life on Earth and have survived. This idea is called **'the survival of the fittest'**.

Excretion

All living things need to get rid of poisonous waste products from their bodies. The removal of these unwanted substances is called **excretion.**

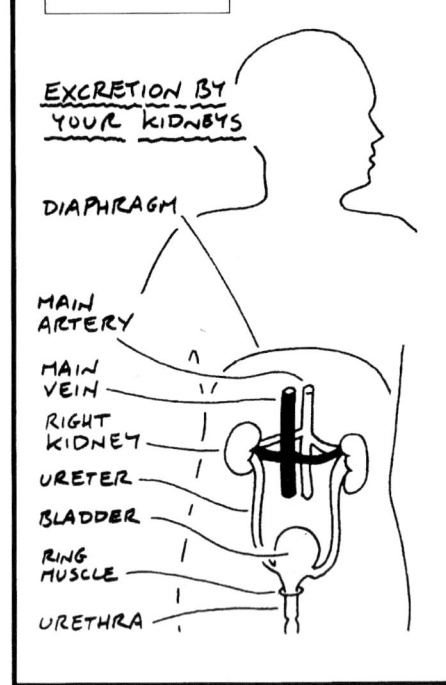

From your lungs, you excrete **water vapour** and **carbon dioxide** - these are both waste products from breathing **(respiration).**

Another waste product is called **urea**. This is produced by your liver when it breaks down certain foods. Urea is filtered out by your kidneys - this, and other unwanted substances, make a liquid called **urine.** Urine drains from the kidneys into the bladder which you empty when you go to the toilet.

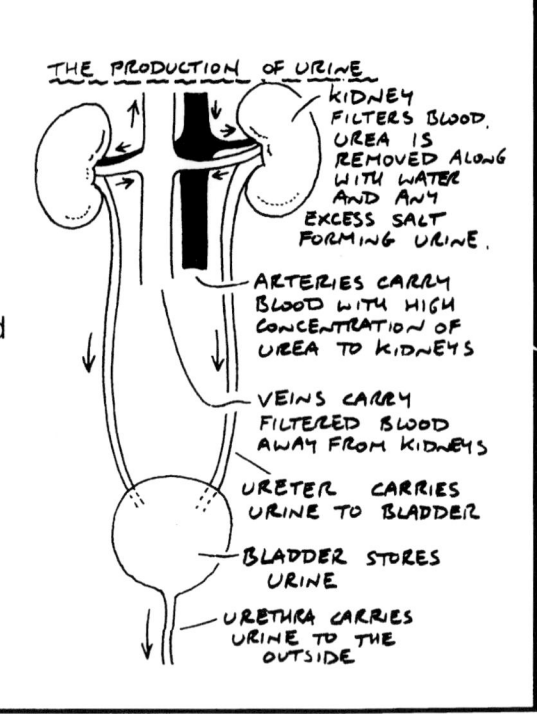

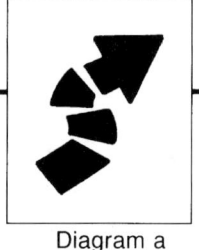

EXCRETION BY YOUR KIDNEYS
- DIAPHRAGM
- MAIN ARTERY
- MAIN VEIN
- RIGHT KIDNEY
- URETER
- BLADDER
- RING MUSCLE
- URETHRA

THE PRODUCTION OF URINE
- KIDNEY FILTERS BLOOD. UREA IS REMOVED ALONG WITH WATER AND ANY EXCESS SALT FORMING URINE.
- ARTERIES CARRY BLOOD WITH HIGH CONCENTRATION OF UREA TO KIDNEYS
- VEINS CARRY FILTERED BLOOD AWAY FROM KIDNEYS
- URETER CARRIES URINE TO BLADDER
- BLADDER STORES URINE
- URETHRA CARRIES URINE TO THE OUTSIDE

Expansion

Most things get bigger when they are heated. They **expand.**
Most **solids** expand as they get hotter, but often by so little that you hardly notice. Sections of concrete or steel used to make bridges or roads expand when they get hot, so gaps are left between them to allow for this expansion. (Diag. a)

Liquids expand as they are heated. Different liquids expand by different amounts. **Alcohol** expands quite a lot when it is heated. This is one of the reasons why it is sometimes used in thermometers. As the liquid in a thermometer warms up, it expands up a very thin tube and shows a higher reading on the temperature scale.

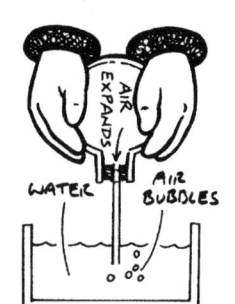

Diagram a
BRIDGE — GAP
ROLLERS
THE ROLLERS LET THE BRIDGE MOVE WHEN IT EXPANDS

Diagram b
WATER — ETHER — WATER BATH — BENZENE — ALCOHOL
HEAT HEAT HEAT HEAT HEAT

Gases expand more than liquids do as they get hotter. If you warm the air in a flask above water as in diagram c, the air expands and pushes the water and then the air in the tube down, making bubbles.

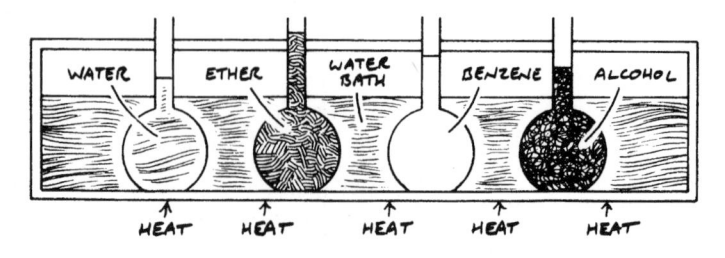

Diagram c
AIR EXPANDS — WATER — AIR BUBBLES

Extinction

All living things gradually change or evolve over a long period of time (thousands of years). They adapt to their environment and then they survive. This is sometimes called **'survival of the fittest'**, and is part of the theory of **evolution**. Humans have changed and adapted over thousands of years and we have survived. Some organisms have not survived - they have become **extinct**.

A bird called the **dodo** became extinct, not long ago, because it was killed by humans.

DODO

Other animals are in danger of becoming extinct. The danger is often made greater by human beings who either hunt and kill certain types of animal, or destroy the **environment** they live in - their **habitat**. Very many types or **species** of living things are becoming extinct every year, partly because of human beings who kill them or ruin their habitat.

BALD EAGLE

There are many **endangered species** on Earth, which need protection if we don't want them to become extinct.

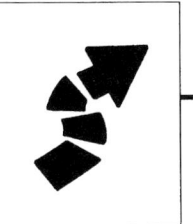

LADY'S SLIPPER (ORCHID)

GIANT PANDA

Fermentation

If apples are left to rot, they start to **ferment**. The sugar inside the apple gradually turns to **alcohol**. Fermentation happens when sugar is 'broken down' by **enzymes** (such as the ones in yeast) to make alcohol and a gas called **carbon dioxide**. Alcohol is a very important chemical for industry.

BARLEY + YEAST → BEER

GRAPES + YEAST ON THE GRAPE SKIN → WINE

The sugar usually comes from plants - fruits or vegetables such as apples, grapes, potatoes or barley. These can be **fermented** to make alcoholic drinks.

POTATO + YEAST → VODKA

Different alcoholic drinks have different amounts of alcohol in them.

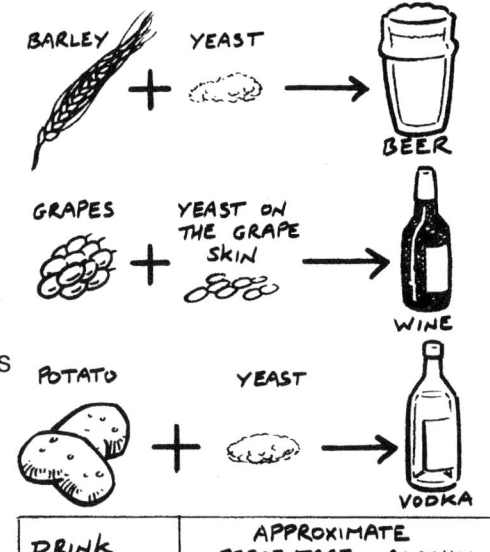

DRINK	APPROXIMATE PERCENTAGE ALCOHOL
BEER	5
WINE	12
SHERRY	25
WHISKY	40

Fertiliser

Plants need **minerals** to help them grow. Some minerals - such as nitrogen, potassium, magnesium and calcium - are present in soil, but they often need to be added to garden and farm soils to keep the soil **fertile.** The materials added to provide these minerals are called **fertilisers.** They dissolve in water and are taken up by plants through their roots.

There are two types:
Organic fertilisers include manure, rotted compost, dried animal blood and bone meal. They take time to act because they must first **decompose** in the soil, but they are very good for the soil.
Inorganic fertilisers are chemicals which act more quickly, but which can be harmful to soil if used over a long period.

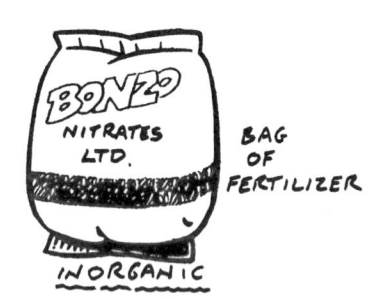

Fertilization

Fertilization happens when a male sex cell (a male **gamete**) joins with a female sex cell (a female gamete). They join together or **fuse** to make a new living thing. The first cell of this new organism is called a **zygote,** which eventually develops into an **embryo.**

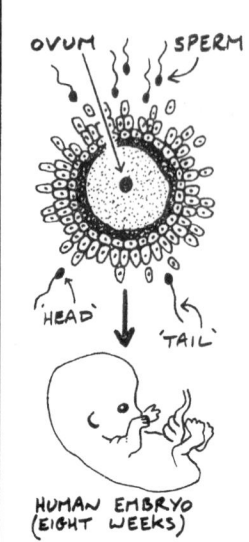

Fertilization can take place inside or outside the body. With **internal** fertilization, **sperms** go inside a female's body and fuse with an **ovum**. This is what happens with human beings. With **external** fertilization, eggs and sperm may meet in water (frogs and most fish do this) or in a 'nest' or hollow.

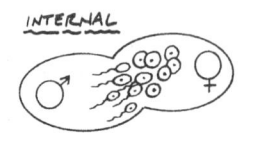

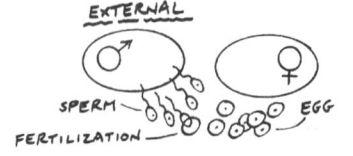

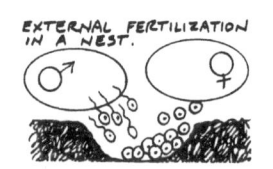

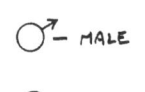

With flowering plants, the male sex cells are in the **pollen**. Pollen grains are carried by birds, insects or the wind to the **stigma** of a flower. Here they make a special tube which carries the male sex cell to the female sex cell (the **ovule**). The male and female sex cells then join or fuse to make a zygote - the first cell of a new plant.

Fission

Fission means 'splitting' - in science, usually of the **atom** or a **cell**.

Large atoms, such as those in **uranium-235**, can be split when a tiny particle called a **neutron** collides with the centre of the atom, which is called the **nucleus**.

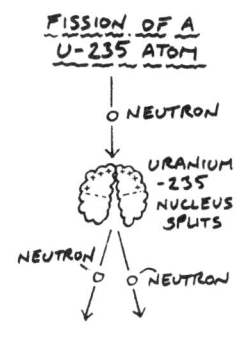

The nucleus of the atom splits into two parts and also sends out more neutrons. If there is enough uranium-235 around then a **chain reaction** may happen.

A chain reaction releases a tremendous amount of **energy**, which can lead to an atomic explosion. This happened with the atomic bombs dropped over Japan in 1945. The energy can also be controlled and harnessed to provide atomic (or **nuclear**) energy for generating electricity in an atomic power station.

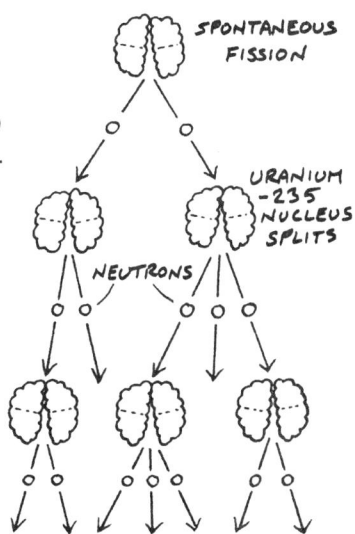

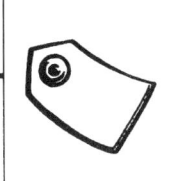

Foetus

A new, developing individual is called an **embryo**. After a human embryo has been in its mother's womb for about 9 weeks (2 months), it is called a **foetus**, and it begins to look like a tiny human being.

The foetus is protected in the womb by a special fluid called the **amniotic fluid**, and it is fed by the **placenta**. This is attached to the foetus by the **umbilical cord**, which is cut and falls off at the 'belly button' after the baby is born.

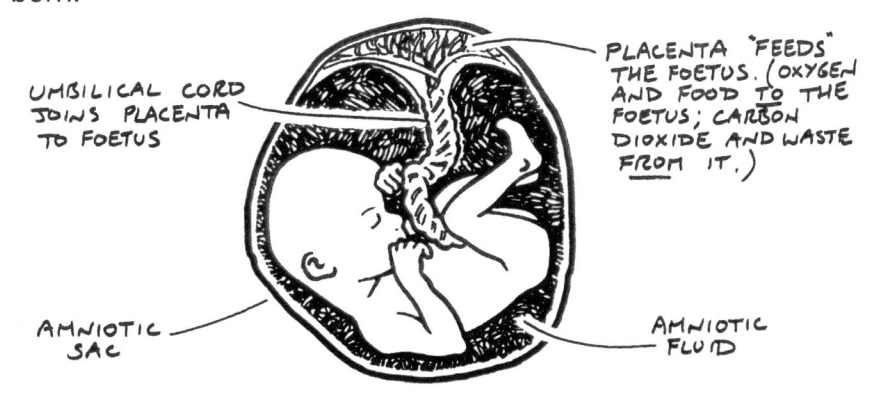

Fossils

Fossils are the remains of dead plants and animals which have been preserved in the ground or in rocks. Sometimes, an organism leaves an imprint or an **impression** of its shape, rather like a footprint in wet concrete that sets hard. Other fossils were made from body parts like bones or shells buried in the ground. **Minerals** which these took in from the soil formed into 'replicas' of the original organism.

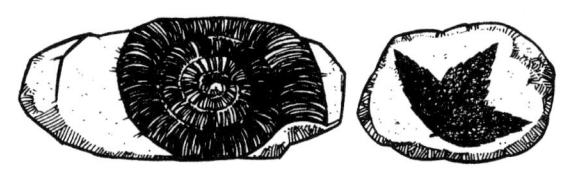

THE FOSSIL RECORD

DIFFERENT GROUPS OF ORGANISMS APPEAR AT DIFFERENT TIMES IN THE EARTH'S HISTORY. THE WIDTH OF EACH "BULGE" REPRESENTS ROUGHLY THE NUMBER OF FOSSILS IN EACH PERIOD.

Fossils give us lots of information about the past, and tell us how organisms have developed or **evolved** over time. The very oldest fossils come from very simple organisms, while newer fossils were made by more complicated organisms. These fossil records support the theory of **evolution**, which says that over millions of years **organisms** have changed and developed.

Frequency

All waves have their own **frequency** - sound waves, water waves, light waves, radio waves and so on.

The frequency of a wave is the number of complete waves made in one second - the wave in this diagram has a frequency of two waves per second or **two hertz (2 Hz)**.

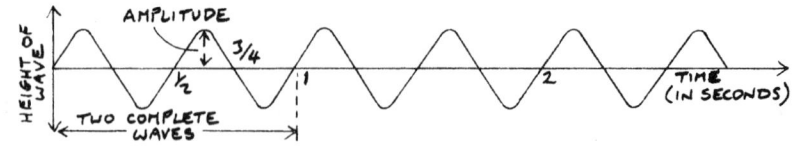

With sound waves, **high-pitched** sounds have a higher frequency than **low-pitched** sounds. A bat makes very high-frequency sounds while flying, which bounce off objects and help it to navigate. Large organ pipes can produce very deep, low-frequency sounds.

The higher the frequency of a wave, the shorter is its **wavelength**. Radio waves have a lower frequency and a longer wavelength than X-rays.

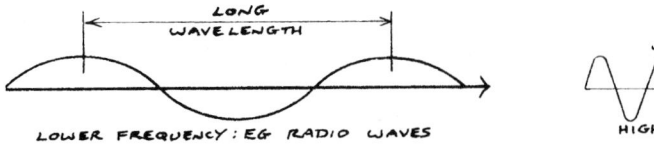

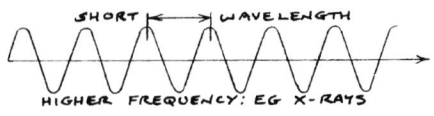

LOWER FREQUENCY: EG RADIO WAVES

HIGHER FREQUENCY: EG X-RAYS

Friction

When you rub your hand up and down on your sleeve, you can feel a force slowing it down. This force between two surfaces is called **friction**.

If two **rough** surfaces rub together, there is a lot of friction.

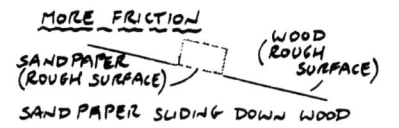

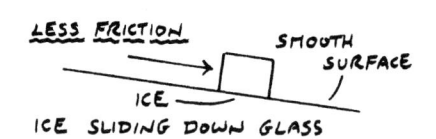

If the surfaces are **smooth** and slippery, however, there is much less friction. But there *can* be high friction forces between smooth sufaces, e.g. bike brakes.

Friction can be useful. It allows us to walk around, to hold on to things and to turn corners. Accidents can happen when there is not enough friction - think of what can happen if you step on a patch of ice or a banana skin!

We often need to reduce the friction between the moving parts of machines, as friction causes wear and tear and also makes things hot. In car engines or bicycles, we do this by putting oil in the space between the parts to **lubricate** them, e.g. on a bicycle chain.

Fuel

Coal, oil, gas, wood, peat and animal dung can all be used as **fuels.** We can burn them to produce heat. The burning of fuels is called **combustion**.

Fossil fuels (coal, oil, and gas) are formed from living things that died millions of years ago. The dead remains were buried under the Earth and were made into the fuels we use today by high pressures and temperatures underground.

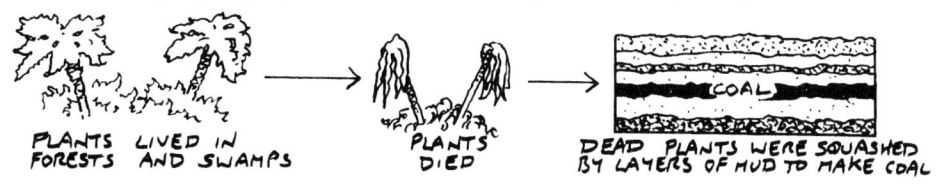

Fossil fuels have been our main supply of energy for over 100 years, but burning them causes problems. They all contain carbon and this makes carbon dioxide gas when the fuel burns. Carbon dioxide in the air causes the 'greenhouse effect' which could lead to **global warming.**
Sulphur in some fuels (e.g. certain types of coal) turns to sulphur dioxide gas when burnt, and this causes **acid rain.**

Fungus

Mushrooms, toadstools, moulds and yeast are all types of fungus. Fungi are neither plants nor animals. Most fungi are 'multicellular', which means they are made up of lots of cells. But yeasts are fungi which consist of one cell only.

YEAST MAGNIFIED x 3000

Yeasts feed on sugar and change it into alcohol and carbon dioxide gas. This process is called **fermentation**, and is used in brewing, wine making and baking.

Mushrooms, moulds and toadstools feed by breaking down or **decomposing** dead organisms like trees or old food.

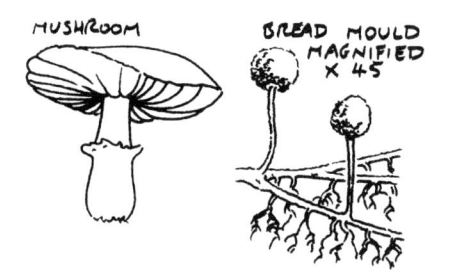

MUSHROOM

BREAD MOULD MAGNIFIED x 45

'Athlete's foot', which makes a person's toes itchy and peeling, is caused by a kind of fungus which lives in the moist, warm conditions inside socks and trainers!

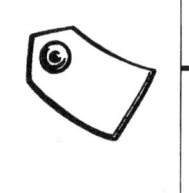

Fuse

A fuse is a short piece of thin wire which gets very hot and melts if too much electric current goes through it.

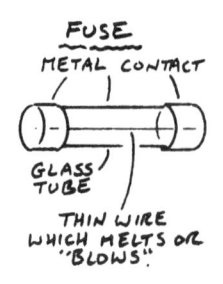

FUSE
METAL CONTACT
GLASS TUBE
THIN WIRE WHICH MELTS OR "BLOWS"

As the fuse melts or 'blows', it leaves a gap in the electrical circuit and the current stops.

Fuses are used for safety. All plugs have their own fuse (next to the 'live' wire), which melts if the current suddenly becomes too large and dangerous.

EARTH WIRE (GREEN/YELLOW)
FUSE
LIVE WIRE (BROWN)
NEUTRAL WIRE (BLUE)
CABLE GRIP

Different electrical devices, such as hair dryers, kettles and cassette players, need different sized fuses because they all take different amounts of current. Current is measured in **amperes,** or **'amps'** or **'A'** for short. Common fuse sizes are 3A, 5A and 13A.

FUSE VALUE	DEVICE
3A	TV, TABLE LAMP, CD-PLAYER, ELECTRIC BLANKET.
5A	VACUUM CLEANER, ELECTRIC IRON, HAIR DRYER, SMALL ELECTRIC FIRE
13A	KETTLE, LARGE FIRE, IMMERSION HEATER

Fusion

The centre or **nucleus** of some atoms can be joined together or combined to make one single nucleus. This joining together is called **nuclear fusion**. It only happens with very small nuclei at very high temperatures (of millions of degrees) and high speeds.

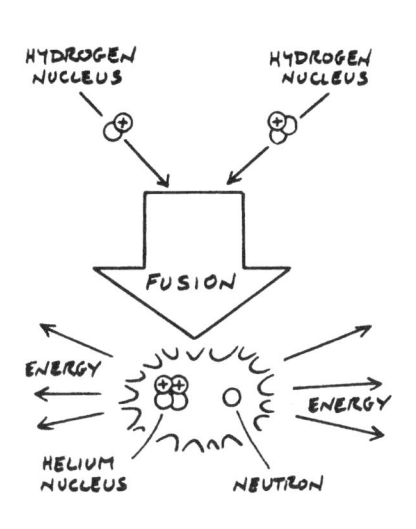

Nuclear fusion produces a tremendous amount of energy. The sun's energy comes from the nuclear fusion of **hydrogen** atoms. On Earth, fusion has been made to happen in the 'hydrogen bomb', in which the splitting or **fission** of larger atoms makes the energy needed to start the joining or **fusion** of smaller atoms. This causes a huge, uncontrolled explosion.

Scientists are now trying to control nuclear fusion so that it can be used to produce energy for peaceful purposes, but this is proving to be very difficult.

Galaxy

A galaxy is a large collection of stars. Some galaxies contain several million stars, others have thousands of millions. Our own galaxy (containing our star, the **Sun**) is called the **Milky Way**. It is about 100,000 light years across – that is the distance that light would travel in 100,000 years!

Our galaxy belongs to what is called a 'local group' of about 30 galaxies. Another member of the group is the Andromeda galaxy which can sometimes be seen in the night sky as a hazy distant patch, about 2.5 million light years away.

Galaxies come in different shapes and sizes – some spiral-shaped, some elliptical, some not a regular shape at all. The Milky Way and Andromeda are spiral galaxies.

Gamete

Most animals and flowering plants make new plants or animals (**reproduce**) sexually. This means that a male cell joins, or fuses with, a female sex cell to make a new single cell called a **zygote**. The zygote then divides many times to produce more cells, making a new individual.

Male and female sex cells are called **gametes**. They are produced by the sex organs of a plant or animal.

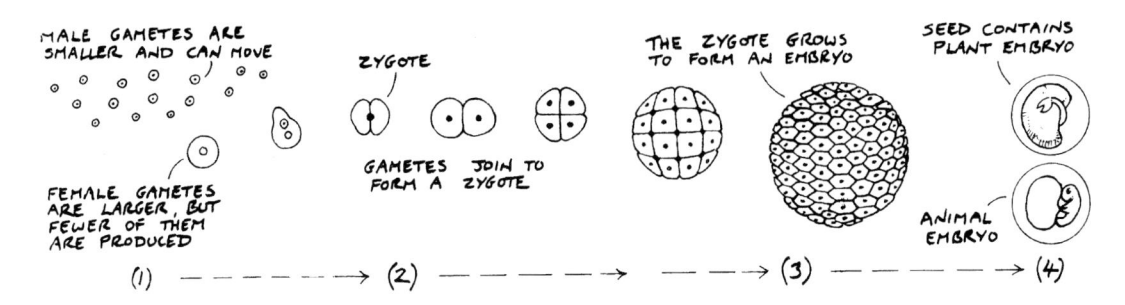

Male gametes in animals are known as **sperm**. They are smaller than female gametes and can move. A female gamete is called an **ovum** (or egg cell) and is usually larger than the male gamete.

Gamma rays

Gamma rays are dangerous rays given off by some radioactive materials. They travel at the speed of light and can only be slowed down by very thick sheets of lead or concrete. Two other types of radiation, alpha and beta, do not penetrate a thick metal sheet as gamma rays do.

Gamma rays are electromagnetic waves like x-rays, but they have a higher frequency and are much more penetrating and dangerous to humans.

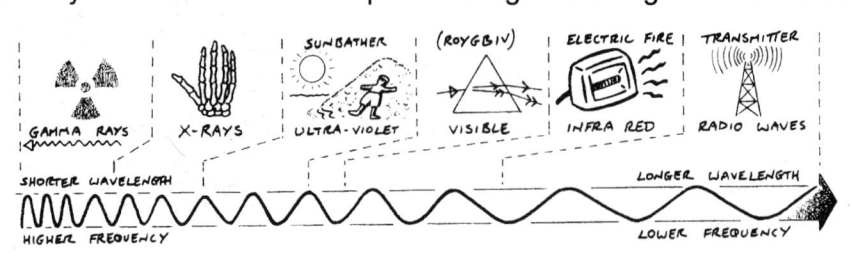

Gases

Solid, **liquid** or **gas** – these are the three conditions or 'states' in which materials can be found. A solid melts to become a liquid; a liquid boils to become a gas. The only material that we see melting and boiling in our everyday lives is water. We see ice (solid) melt to become water (liquid) and then sometimes boil to become steam (gas).

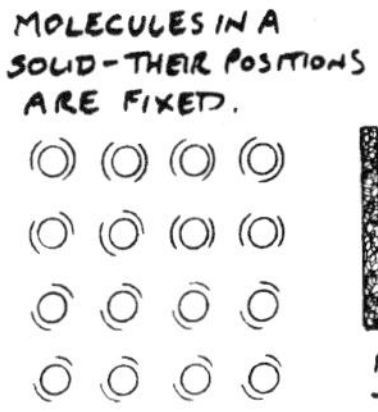

MOLECULES IN A SOLID – THEIR POSITIONS ARE FIXED.

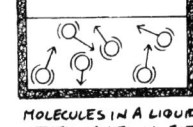

MOLECULES IN A LIQUID – THEY CAN FLOW, BUT THEY ARE NOT FREE

Air is a gas, or really a mixture of gases like oxygen, nitrogen and carbon dioxide. Some gases, like air, are invisible – but others have a colour. Chlorine is a greeny yellow colour while bromine is brown (both are poisonous gases).

All gases need to be kept in a container, otherwise they spread out or **diffuse** to take up as much space as they can. This is because the particles that make up a gas are totally free. In a liquid they can flow around but are not completely free – liquids need a container but not on all sides. The particles in a solid are more or less fixed in position – solids don't need to be trapped by a container.

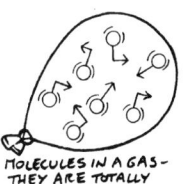

MOLECULES IN A GAS – THEY ARE TOTALLY FREE TO MOVE AND SPREAD OUT TO FILL THE CONTAINER. THEY ARE IN COMPLETELY.

WATER IN THREE STATES

SOLID (ICE) — SOLIDS HAVE A FIXED VOLUME AND SHAPE.

LIQUID (WATER) — LIQUIDS HAVE A FIXED VOLUME BUT TAKE THE SHAPE OF THE CONTAINER.

GAS (STEAM) — GASES DON'T HAVE A FIXED VOLUME – THEY SPREAD OUT TO OCCUPY ALL THE SPACE AVAILABLE.

Generators

Generators are devices used to make electricity. They can be very small, like the dynamos on a bicycle wheel or very large, like the huge generators in power stations.

All generators, large or small, work by having a coil of wire moving inside a magnet. As the wire moves through the magnetic field an electric current is made or induced.

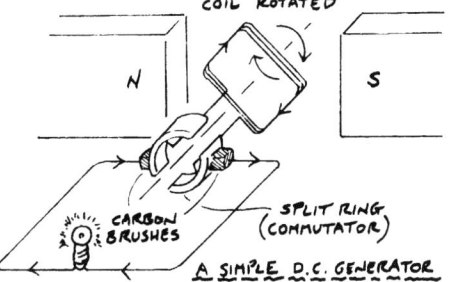

A SIMPLE D.C. GENERATOR

Some generators make alternating current (AC), some make direct current (DC)

The faster the coil of wire turns, the stronger the current. The coil can be made to move by a bike wheel or by steam pressure forcing it round (this happens in power stations).

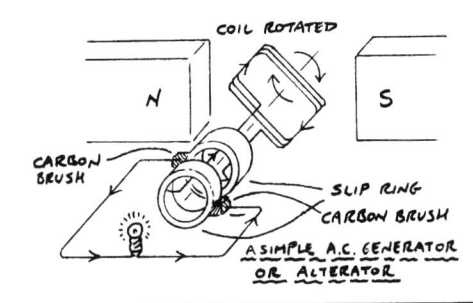

A SIMPLE A.C. GENERATOR OR ALTERNATOR

Genes

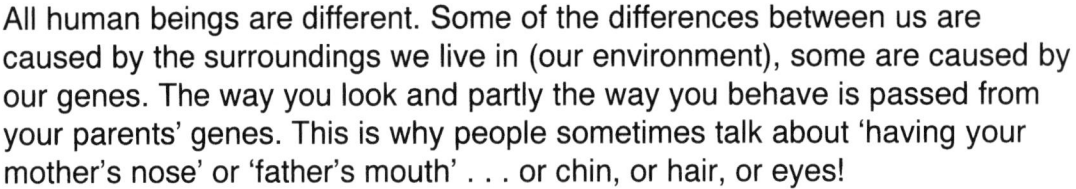

All human beings are different. Some of the differences between us are caused by the surroundings we live in (our environment), some are caused by our genes. The way you look and partly the way you behave is passed from your parents' genes. This is why people sometimes talk about 'having your mother's nose' or 'father's mouth' . . . or chin, or hair, or eyes!

A gene is like a piece of code or information which partly controls the way we are. A sperm contains a set of genes from a male parent, the ovum has the genes from the female. At fertilisation these two sets of genes come together – they then control the development of the fertilised ovum into an embryo, which develops into a baby.

Genes are like orders or instructions – some are stronger than others. If genes for brown eyes meet genes for blue eyes, the child will be brown-eyed. The gene or instruction for brown eye colour is stronger – it is the **dominant gene**.

Some diseases are passed on from parents to a child through genes. The study of the way genes work and how they pass from one generation to the next is called **genetics**.

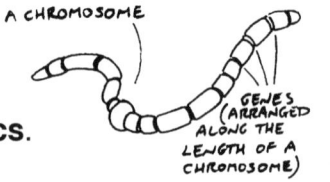

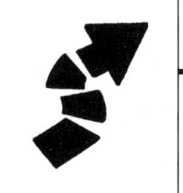

Germination

The germination of a seed is the start of a plant's life cycle, when it 'sprouts' and seems to come to life.

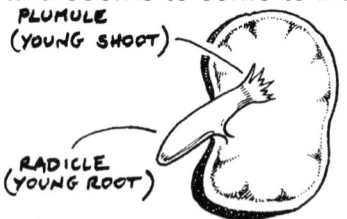

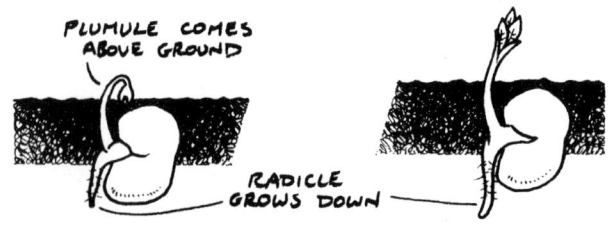

To germinate and develop into a plant, a seed needs water, oxygen and warmth. It absorbs water, swells up and the seed coat bursts open. As it sprouts, a **radicle** (a young root) and a **plumule** (a young shoot) appear.

Seeds often germinate in soil, but they can germinate without it. For example, you can grow cress seeds on damp cotton wool or kitchen towel.

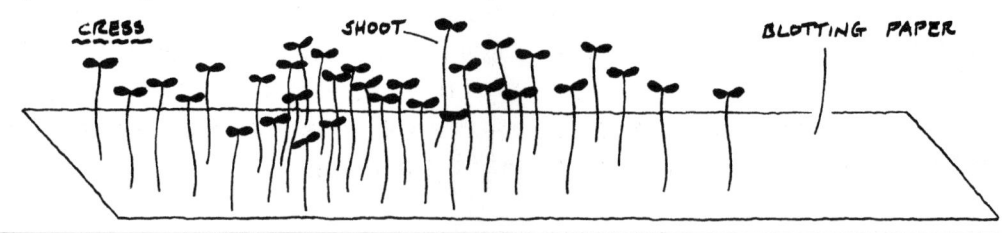

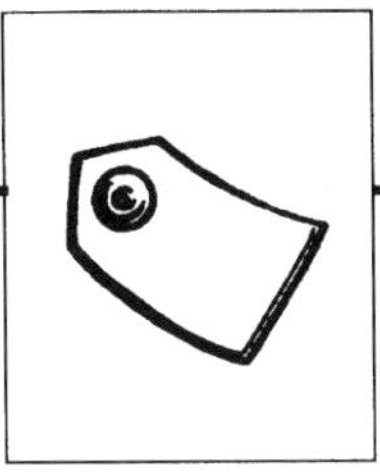

Glucose

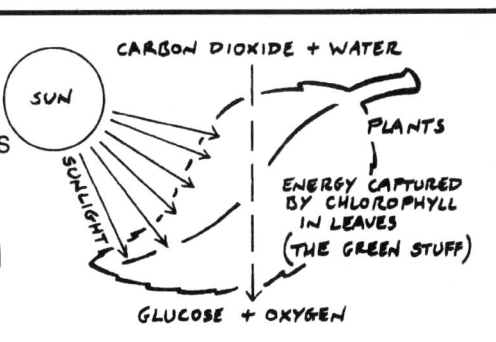

Glucose is a **sugar** which is found in honey, sweet fruits and other kinds of food that we eat, e.g. carbohydrates. It is made by **photosynthesis** in plants.

Carbon dioxide and water are converted into glucose and oxygen.

Glucose provides us with energy. **Enzymes** in our cells break down the glucose molecules to release the energy stored within them. This is called **respiration**. Glucose and oxygen react together to release energy and make carbon dioxide and water.

$$\text{Glucose + oxygen} \xrightarrow{\text{enzymes}} \text{carbon dioxide + water (+ energy)}$$

Glucose is also used in **fermentation**. Yeasts convert sugar into alcohol and carbon dioxide.

Gravity

Gravity is the force which makes things fall, and it acts on you and everything else on Earth all the time. The pull of the Earth's gravity on you is called your **weight**. The pull of gravity is measured in **newtons** ('N' for short).

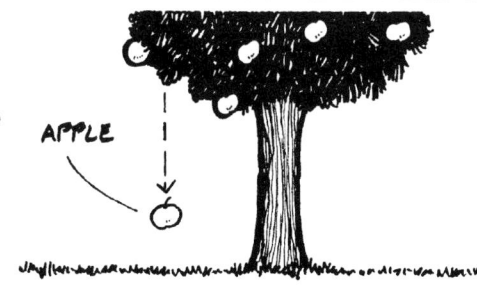

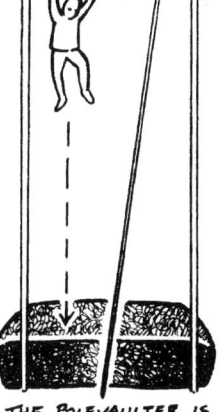

THE POLEVAULTER IS PULLED DOWN TO THE GROUND BY THE FORCE OF GRAVITY.

While your mass is always the same, your weight changes with where you are. The Moon is much smaller than the Earth – and you weigh far less there. Its gravitational pull is much weaker than the Earth's.

Sir Isaac Newton was the first to realise that gravity acts everywhere in the universe - it is a 'Universal force'. All objects attract one another with the pull of gravity, but most objects are too small for their pull to be noticed. Extremely large objects, like stars and planets, have a pull of gravity which is strong enough to affect other things. For example, the Moon's gravity causes the tides in the oceans on the Earth's surface.

Habitat

This is the place or area where an animal or a plant lives. Deserts, jungles, rivers, ponds, towns and cities, woods and forests are all habitats. Different animals and plants live and survive in different habitats, e.g. penguins survive in the frozen conditions of the Antarctic.

A **community** is the group of living things which live in a habitat. The habitat and the community of organisms that live in it together make up an **ecosystem**. For example, a rock pool is a habitat with a community which might include seaweed, crabs, starfish and sea anemones.

All of the organisms in a habitat depend on each other for their survival. They are linked by **food chains**, which together form a **food web**.

Heart

Your heart is the organ that pumps the blood around your body. It is about the same size as a clenched fist. The human heart is mostly made of special muscle called cardiac muscle, and contracts about 70 times each minute.

Your heart has four compartments. At the top are the right atrium and the left atrium. These are where the blood enters the heart.

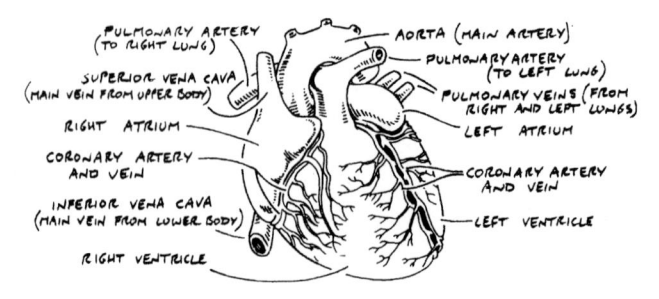

The lower compartments are the left and right ventricles, with thicker walls. Inside the heart are special one-way valves which allow blood to pass from the atrium to the muscular ventricle beneath. The right ventricle pumps blood to the lungs to pick up oxygen. The left ventricle pumps oxygen-rich blood to all other parts of the body.

Helium

Helium is a very light gas, almost as light as hydrogen (the lightest gas). It is used to fill airships and balloons which float in the earth's atmoshere.

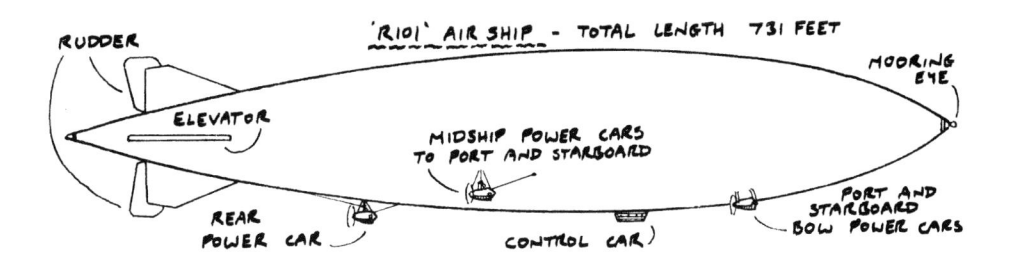

Helium is **less dense** than air, which is why helium-filled balloons float. Hydrogen was once used to fill airships, but in 1937 there was a terrible disaster in the United States when the German airship *Hindenburg* caught fire and nearly 50 people were burned to death. Nowadays, helium is used instead for filling airships because it does not burn (it is **non-flammable**). In fact, helium does not take part in any chemical reactions at all – it is called a 'noble gas'.

Helium is also used in liquid form to keep things at very low temperatures – it becomes a liquid at minus 269°C.

Herbivore

A herbivore is an animal which eats only plants. It does not eat other animals, as **carnivores** or **omnivores** do. Cattle, sheep, horses, rabbits, greenflies and mice are all herbivores.

Herbivores are sometimes eaten by carnivores, e.g. greenflies are eaten by ladybirds. These carnivores are then eaten by other carnivores (e.g. ladybirds are eaten by birds). This is called a **food chain**.

Sunlight provides all the energy needed to keep living things on Earth alive – plants convert this into food which herbivores eat. Herbivores are also called **primary consumers**, because they only eat plants. **Secondary consumers** are carnivores like foxes, dogs, hawks and cats, which eat herbivores.

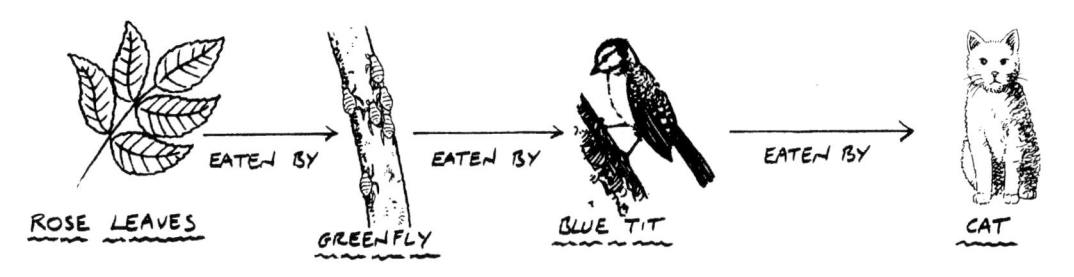

Hormone

Our bodies are controlled by two systems – the **nervous system** and the **hormone system**. Hormones are chemicals carried by the blood and they control many different activities. For example, if you are frightened your body produces a hormone called adrenalin – this makes your heart beat faster and blood rush to your muscles.

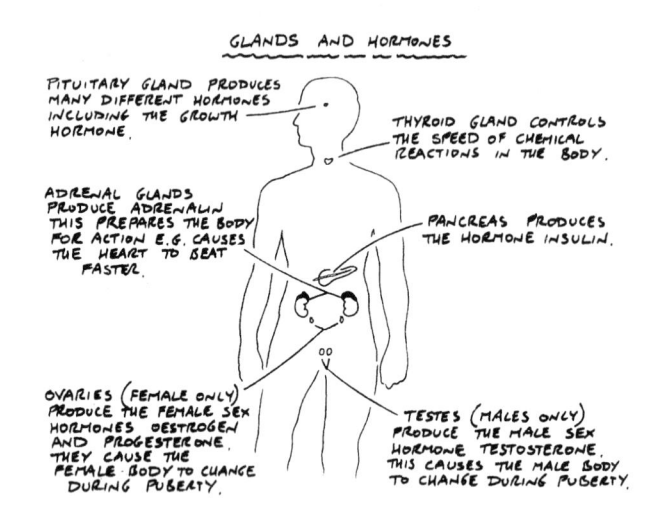

GLANDS AND HORMONES

PITUITARY GLAND PRODUCES MANY DIFFERENT HORMONES INCLUDING THE GROWTH HORMONE.

THYROID GLAND CONTROLS THE SPEED OF CHEMICAL REACTIONS IN THE BODY.

ADRENAL GLANDS PRODUCE ADRENALIN. THIS PREPARES THE BODY FOR ACTION E.G. CAUSES THE HEART TO BEAT FASTER.

PANCREAS PRODUCES THE HORMONE INSULIN.

OVARIES (FEMALE ONLY) PRODUCE THE FEMALE SEX HORMONES OESTROGEN AND PROGESTERONE. THEY CAUSE THE FEMALE BODY TO CHANGE DURING PUBERTY.

TESTES (MALES ONLY) PRODUCE THE MALE SEX HORMONE TESTOSTERONE. THIS CAUSES THE MALE BODY TO CHANGE DURING PUBERTY.

Other hormones include insulin, which controls sugar in the body, growth hormones which affect your height, and sex hormones which cause the body to change during puberty.

Hormones are produced in tiny amounts by **endocrine glands** in different parts of the body.

Plants also produce hormones.

Hydrogen

Hydrogen is the 'lightest' (least dense) gas that exists. It is also the most plentiful element in the Universe.

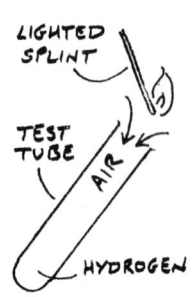

Hydrogen is invisible and has no taste or smell, but it is highly flammable. A test tube full of hydrogen makes a small 'pop' if a light is brought near it.

When hydrogen (H) burns in air it combines with oxygen (O) to make water (H_2O) – this means that it is a very clean fuel, but very dangerous! It was used in airships before the Zeppelin disaster of 1937 when an airship filled with hydrogen caught fire and many people died.

LIGHTED SPLINT
TEST TUBE
AIR
HYDROGEN
BURNING HYDROGEN

Hydrogen can be made by adding an acid (like dilute hydrochloric acid) to a metal such as zinc. Hydrogen bubbles are made and can be collected in a jar.

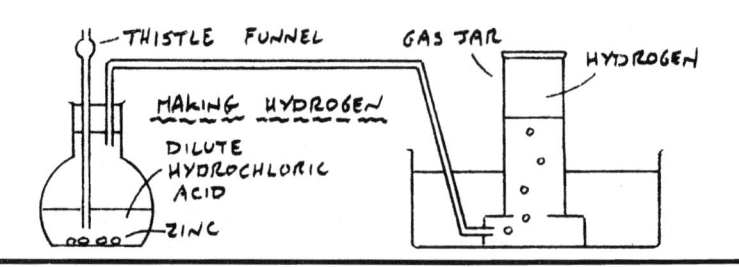

THISTLE FUNNEL   GAS JAR   HYDROGEN
MAKING HYDROGEN
DILUTE HYDROCHLORIC ACID
ZINC

Image

When you look at yourself in a mirror you see an image - your 'mirror image'. In a straight or **plane** mirror, your image is the same size as you are:

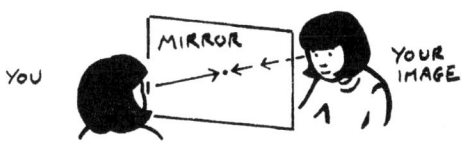

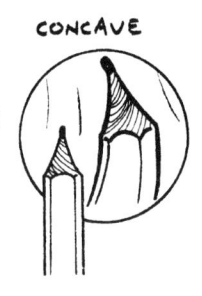

With curved mirrors, the image may be smaller than the object in front of it, or it could be larger. A mirror which curves inwards, a **concave** mirror, makes an image which is larger than the object in front of it.

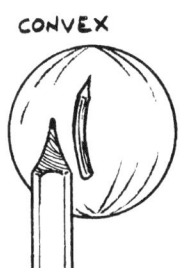

But a mirror which curves outwards, a **convex** mirror, makes an image which is smaller than the object in front of it.

A lens which bulges on both sides (a convex lens) could be used to focus an image of a distant object (like a tree, or the Sun during an eclipse) onto a screen. The image would be upside down.

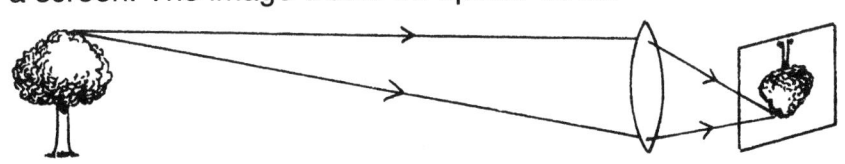

Immunity

Your body needs to defend itself against harmful organisms like some bacteria, viruses and fungi. Your first line of defence is your skin. After that, your body's defence or resistance to infection and disease depends upon your **immunity**.

Natural immunity comes partly from white blood cells. These sometimes surround harmful germs and 'digest' them, but they also produce special chemicals called **antibodies** which attack bacteria and viruses.

You can develop **artificial immunity** from an injection called a **vaccination** (or **inoculation**), which puts a tiny amount of a germ into your body. Your blood responds by making antibodies to kill it. This builds up your immunity to the germ, so that if you ever catch a large dose of it, your body already has the antibodies it needs to defend you.

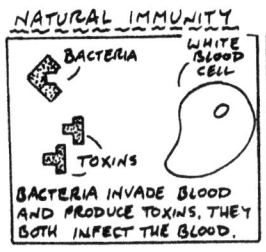

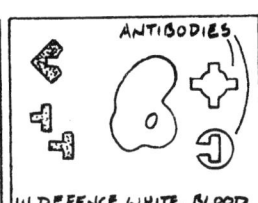

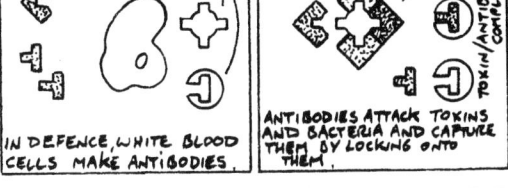

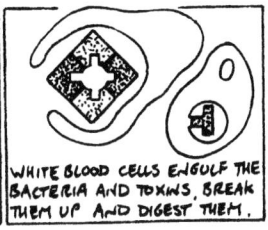

Indicator

An indicator is a chemical substance which turns different colours in acids and alkalis. Many indicators come from natural dyes in plants, for example red cabbage.

blue litmus paper turns red in acid solution

red litmus paper turns blue in alkali solution

A common indicator is called **litmus**. Blue litmus turns red if acid is added to it. Red litmus turns blue if an alkali is added to it.

Universal indicator is made from a mixture of different indicators. Its colour is compared with a diagram which shows the colour it turns with acid, alkali or neutral (in-between) substances.

A number called the **pH** is given to each point on the scale. Strong acids have a pH of 1. At the other extreme, a pH of 14 indicates a strong alkali. Pure water is neutral at pH7.

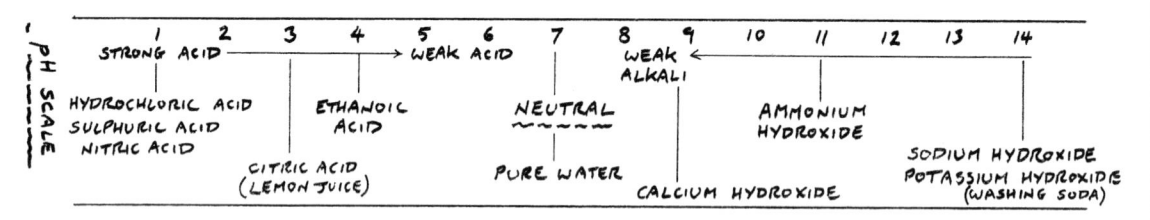

pH SCALE: 1 2 3 4 5 6 7 8 9 10 11 12 13 14
STRONG ACID → WEAK ACID — WEAK ALKALI ←
HYDROCHLORIC ACID, SULPHURIC ACID, NITRIC ACID
ETHANOIC ACID
CITRIC ACID (LEMON JUICE)
NEUTRAL — PURE WATER
AMMONIUM HYDROXIDE
CALCIUM HYDROXIDE
SODIUM HYDROXIDE, POTASSIUM HYDROXIDE (WASHING SODA)

Infrared

Infrared rays, or 'heat waves', are types of wave in the family called the **electromagnetic spectrum**.

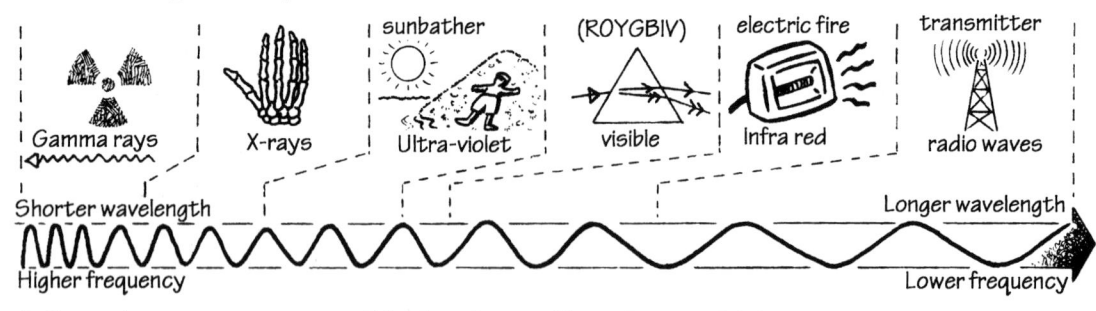

Gamma rays | X-rays | Ultra-violet (sunbather) | visible (ROYGBIV) | Infra red (electric fire) | radio waves (transmitter)

Shorter wavelength → Longer wavelength
Higher frequency → Lower frequency

Infrared rays are responsible for the **radiant heat** which we can feel, but not see, if we stand in front of an electric fire or sit in the sun. They travel at the speed of light and can pass through empty space (a vacuum).

They are given off by all hot objects and they travel through fog, dust or haze better than light does. Thus, although they are invisible, they can be used for 'seeing' objects which do not transmit light. For example, people give out infrared radiation, so they can be picked up using **infrared cameras**, which shows where the heat waves have come from.

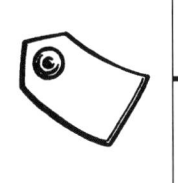

Insulation

Insulation prevents heat, sound or electricity from flowing freely. Many mammals or birds are insulated by a layer of fat beneath the skin or fur, or by feathers which trap a layer of air next to their skin. This helps to keep the heat energy inside their bodies in cold conditions.

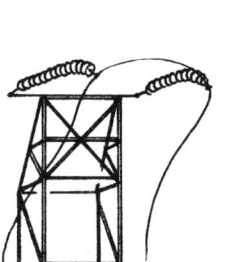

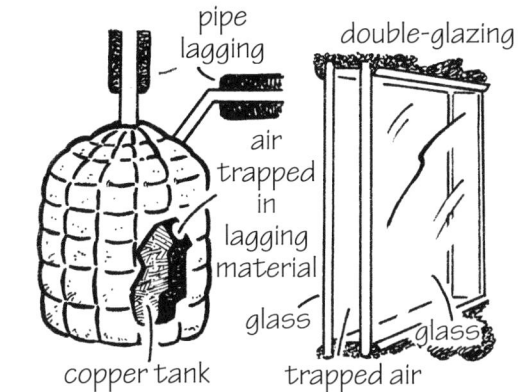

Our homes have **thermal insulation** such as cavity walls, the lagging around hot water tanks and double glazing (air trapped between two sheets of glass).

Electrical insulation helps to make electricity safe. They allow us to confine the flow of electric current to where it is wanted. Plastic, rubber, porcelain and glass are all good **insulators**.

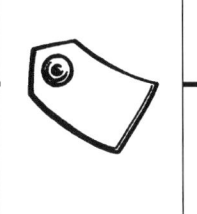

Invertebrate

Animals can be divided up, or **classified**, into two types: animals with backbones (**vertebrates**) and animals without backbones (**invertebrates**).

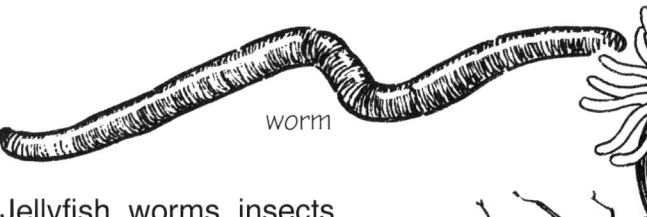

Jellyfish, worms, insects, centipedes, mussels and sea anemones are all examples of invertebrates.

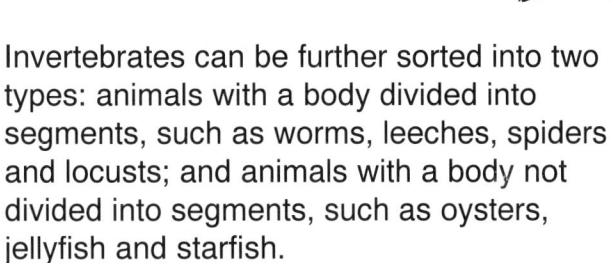

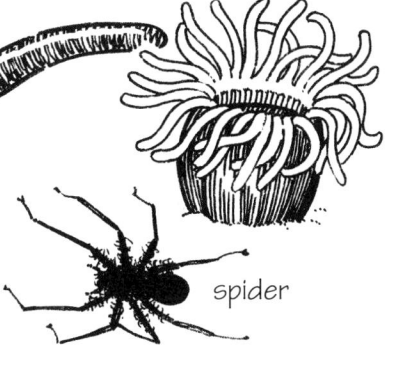

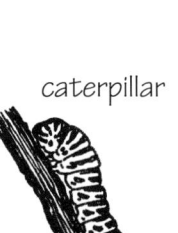

Invertebrates can be further sorted into two types: animals with a body divided into segments, such as worms, leeches, spiders and locusts; and animals with a body not divided into segments, such as oysters, jellyfish and starfish.

Joule

A joule (J) is the unit for measuring **energy**. You need energy to do work so the joule is also the unit for measuring work in science. If you need a **force** of one newton from your hand to lift an apple and you move it up through a distance of one metre, you have done one joule of work (and used one joule of energy).

A weightlifter raising a heavy weight of 500 newtons up by 2 metres does 1,000 joules of work (500 x 2 = 1,000).

1,000 joules is one **kilojoule** (kJ). This is a good unit for measuring the energy values of different foods.

fish & chips about 3,000kJ
one pint of milk 1,600kJ
one apple 170kJ
one carrot 85kJ
one peanut 25kJ
one cherry 10 kJ

The energy you use while doing different activities – like walking, swimming or sleeping – can also be measured in kilojoules. This table shows roughly the energy used by a young person in one minute doing different activities:

activity	energy used each minute
sleeping	4KJ
watching TV	6KJ
walking	14KJ
running	25KJ
running and jumping	30KJ
swimming	32KJ

Energy used by a teenager in one minute

Key

A key is a scheme or diagram which can be used to identify animals or plants. At each stage of the key you are asked a question – or given a statement – about the main features or characteristics of living things. Answering one question, then moving on to the next one, leads you to the name of the animal or plant.

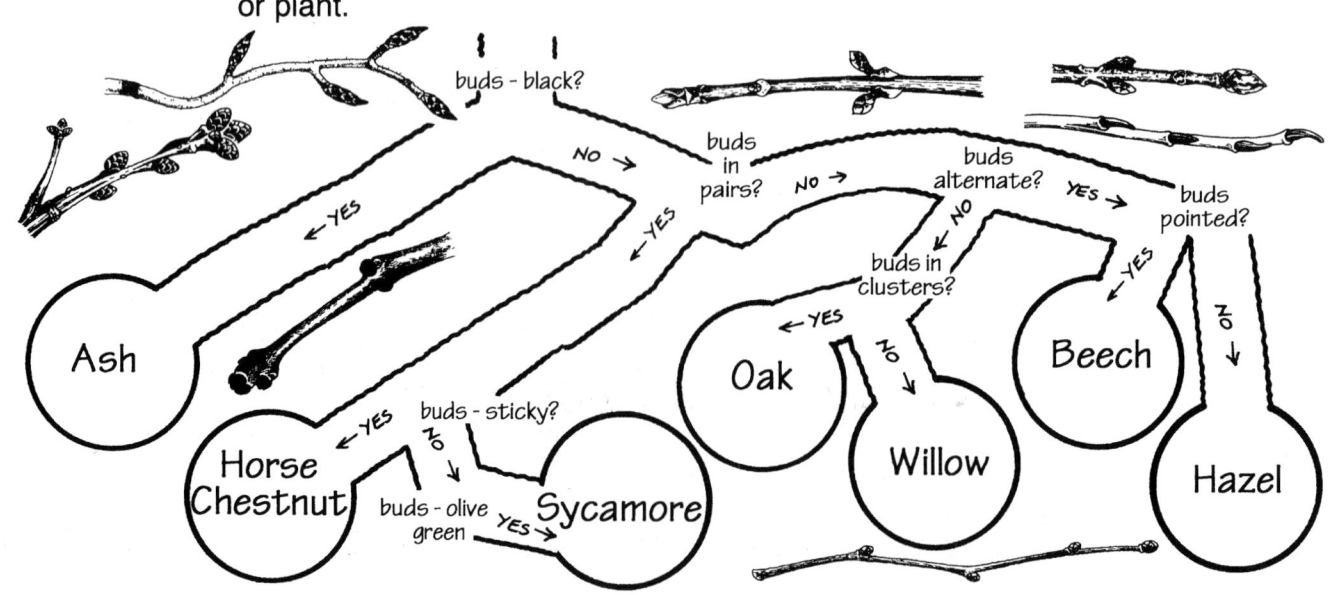

buds - black?
buds in pairs?
buds alternate?
buds pointed?
buds in clusters?
buds - sticky?
buds - olive green

Ash, Horse Chestnut, Sycamore, Oak, Willow, Beech, Hazel

Kidney

Kidneys are a pair of organs inside the back of the body, just above the waist. They control the amount of water in the body and remove unwanted soluble substances.

Kidneys clean blood by **filtering** it. Each kidney contains millions of tiny tubes called **nephrons** which filter the blood. The filtering produces concentrated **urine**. This drains out of the kidneys into the bladder, which is emptied when people urinate.

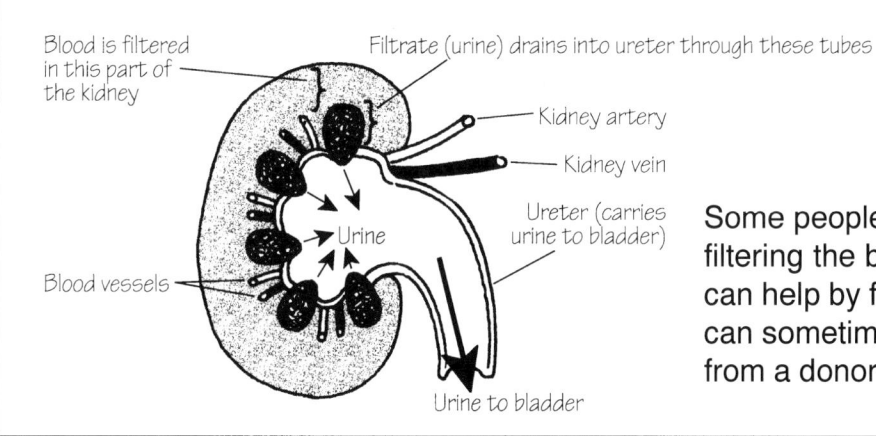

Blood is filtered in this part of the kidney
Filtrate (urine) drains into ureter through these tubes
Kidney artery
Kidney vein
Ureter (carries urine to bladder)
Urine
Blood vessels
Urine to bladder

Some people's kidneys are not very good at filtering the blood – a kidney dialysis machine can help by filtering it for them. A faulty kidney can sometimes be replaced by a healthy one from a donor – this is a kidney transplant.

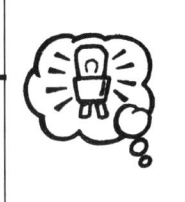

Kilogram

A kilogram (kg) is the unit used for measuring mass (**mass** is the amount of material in an object – it is not the same as weight which is the force of gravity on an object). Ten apples will have a mass of about 1kg; their weight on Earth will be about ten **newtons**. On the moon they will weigh a lot less, but their mass always stays the same – unless somebody takes a bite out of them!

A mass of one kilogram weighs about 10 newtons on Earth

Car
1 tonne = 1000 kg
1m Lead
$1m^3$ of lead = 11,400 kg

Sugar 1kg
1m Water
$1m^3$ of water = 1,000kg

Ten apples weigh about 10 newtons on Earth

Larger masses can be measured in **tonnes** or 1,000 kilograms. Smaller masses can be measured in grams (1,000g = 1kg).

Kinetic

'Kinetic' means movement (from the Greek word 'kinesis'). Every moving object has **kinetic energy** – its energy of movement. Kinetic energy can be measured in **joules** (J).

People jogging, walking or riding a bicycle all have kinetic energy. A moving car has lots of kinetic energy. The faster it moves, the more kinetic energy it has. A heavier car has more kinetic energy than a lighter one going at the same speed.

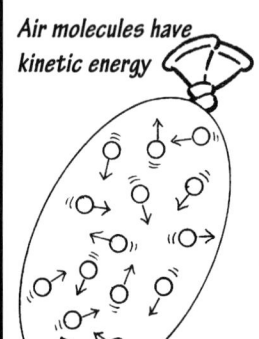

Air molecules have kinetic energy

An object swinging backwards and forwards has most kinetic energy in the middle of the swing, when it is moving fastest (e.g. a pendulum).

The molecules inside any material (e.g. the air molecules inside a balloon) are moving and so have kinetic energy. The hotter they get, the more kinetic energy they have.

Larva

Some baby animals, such as human babies, look (more or less) like their parents. Many baby animals, however, do not look at all like the adult animal – they change their appearance entirely. A baby creature like this is called a **larva** (plural **larvae**). A tadpole is the larva of a frog. A maggot is the larva of a fly. Some larvae are called 'nymphs'.

When a larva develops into an adult, the process is called **metamorphosis**.

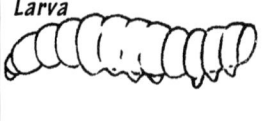

Larva

complete metamorphosis (4 stages) — adult / pupa / butterfly / egg / larva

LIFE CYCLES

incomplete metamorphosis (3 stages) — adult grasshopper / egg / nymph

Barnacle larva (much magnified)

There are good reasons why some animals have a larva. A larva is very small compared with the adult, and adults can produce very large numbers of them so that some are likely to survive.
Also, larvae in the sea may be carried hundreds of miles by ocean currents, helping to spread the species around the world.

Lens

There are two common types of lens – **convex** and **concave**.
A convex lens brings a parallel beam of light to a focus. It is a **converging** lens.
A concave lens makes a parallel beam spread out or **diverge**.

Your eye uses a convex lens to help focus light from objects you are looking at onto the back of the eye (the **retina**).

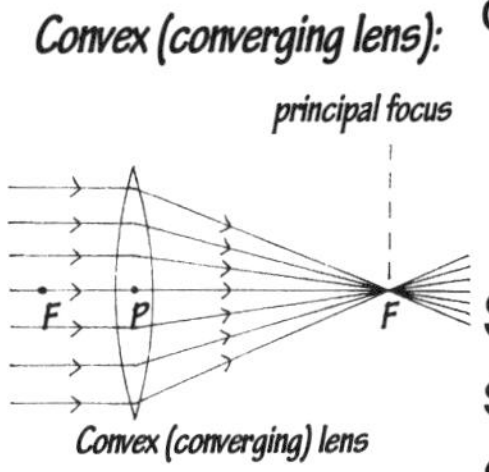

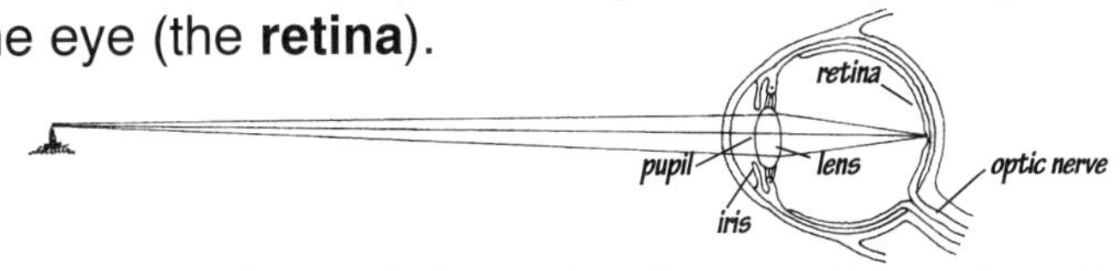

Convex (converging lens):
principal focus
Convex (converging) lens

Some people's eyes cannot focus light perfectly onto the retina. If they are short-sighted, light is focused in front of the retina. They need spectacles with a diverging lens to correct this.

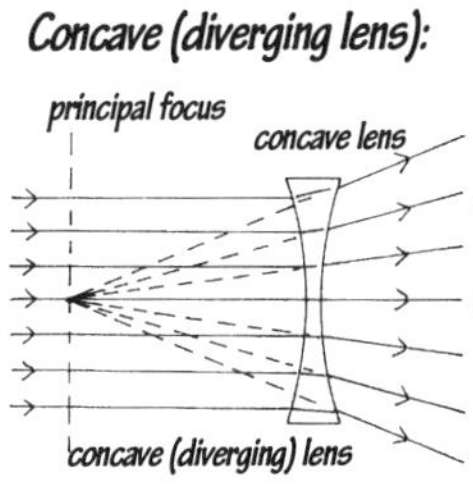

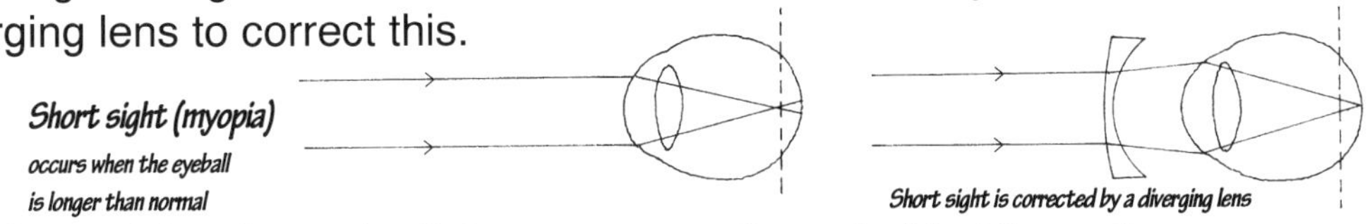

Concave (diverging lens):
principal focus
concave lens
concave (diverging) lens

Short sight (myopia) occurs when the eyeball is longer than normal

Short sight is corrected by a diverging lens

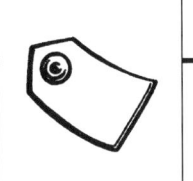

With long-sighted people, light comes to a focus behind the retina, so a converging lens is needed to focus it properly.

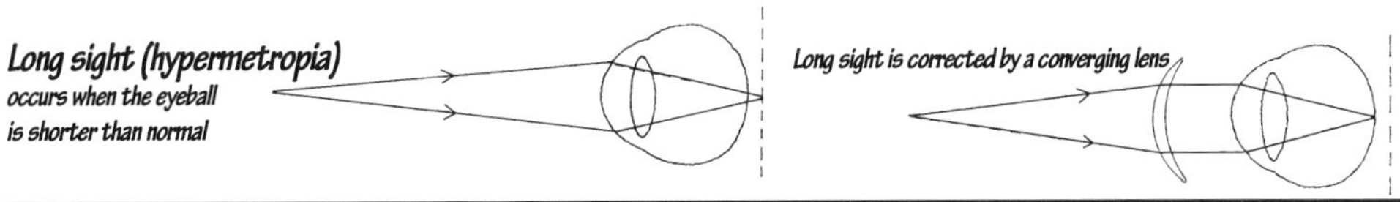

Long sight (hypermetropia) occurs when the eyeball is shorter than normal

Long sight is corrected by a converging lens

Lever

A lever can be used to lift a much heavier weight than could be lifted without one.

Every lever needs a **pivot** (sometimes called a **fulcrum**). Here are some more levers:

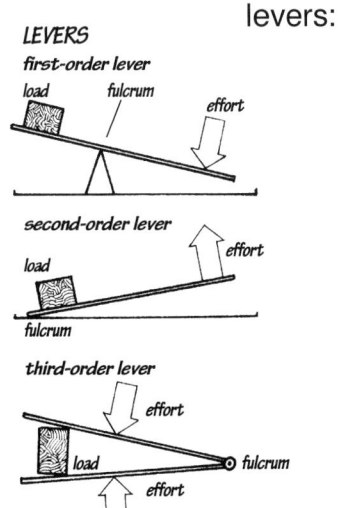

LEVERS
first-order lever
load   fulcrum   effort

second-order lever
load   effort
fulcrum

third-order lever
effort
load   fulcrum
effort

pliers
pivot

raising a heavy wheel barrow
force
pivot
weight

effort
load   pivot

claw hammer
pivot

A first-order lever multiplies your effort. A long lever will move a very heavy weight with very little effort.

Ligaments

Ligaments are tough, stretchy fibres in your body. There are different types.

Some hold bones together, such as the joints in the elbows and knees.

In the eye, ligaments are attached to the eye lens. The lens can be made longer and thinner, or fatter and shorter. This change allows you to see objects close up (fatter lens) or farther away (thinner lens).

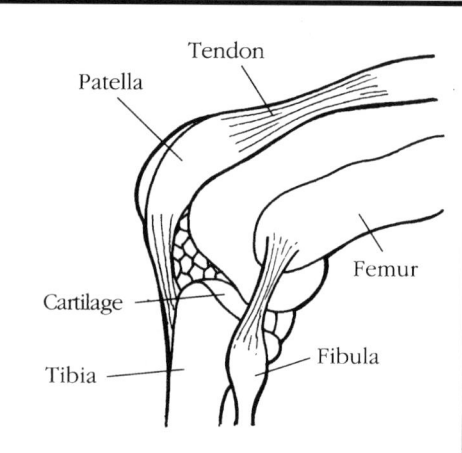

A woman's body has ligaments which attach the ovaries to the uterus.

We have ligaments holding the roots of our teeth in place in our mouth. These are called 'periodontal ligaments' – they join the tooth to the jaw bone.

Liquid

In everyday life, we see many liquids: milk, water, petrol and oil. Liquids are in one of the three 'states' of matter – the others are **solids** and **gases**. Water is solid when it is ice. At boiling point it becomes a gas.

Liquids like water take the shape of the container you keep them in.

The particles in a solid are arranged in a regular pattern – when they are heated this pattern starts to break down as the solid melts.

All the time, particles from a liquid are escaping into the air – they are evaporating. If you heat a liquid up, all the particles in the liquid break away and turn to gas. Eventually, at boiling point, the liquid becomes a gas. Water boils at 100°C.

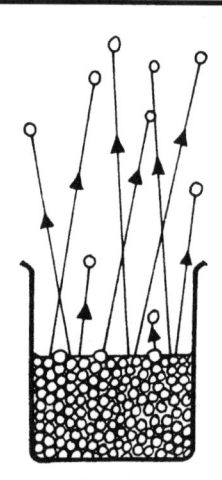

Boiling water molecules

Loudness

The loudness of a sound depends partly on the size or **amplitude** of its sound wave. Noisy sounds have a large **amplitude**, quieter ones have a smaller amplitude.

The loudness of sounds is measured in **decibels** (dB for short). Quiet sounds, like a clock ticking or a person whispering, have a loudness of about 20 to 30 decibels. A noisy factory or heavy traffic are about 90 to 100 decibels. Concorde can be around 200 decibels. The decibel scale is different from many other scales. 20**dB** is more than twice as loud as 10**dB**.

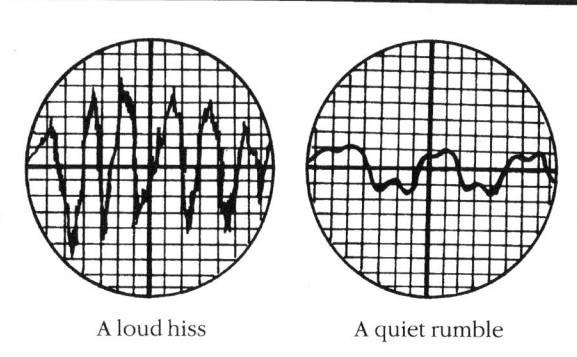

A loud hiss   A quiet rumble

Luminous

Any object that gives out its own light is luminous. The Sun is luminous – it produces its own light and heat rays from nuclear reactions within. The Moon is not luminous – it does not produce its own light. We can see it in the sky because it reflects light from the Sun.

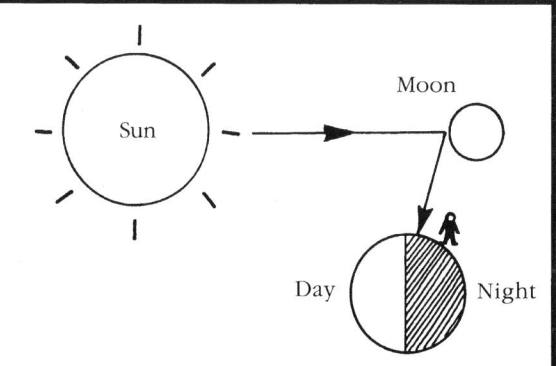

Most objects we see, such as trees, cars, buildings or other humans, are not luminous. We see them because they reflect light from a luminous object. In 'daylight', they are reflecting light from the Sun – at night time we see things if they reflect light from a streetlight, a lighthouse, or a torch, or even light already reflected from the Moon.

Artificial light from light bulbs is often given out by a very hot wire or filament which glows when electricity goes through it. **Natural light** can come from the Sun or from other stars – they are luminous because they are very hot. The hotter the star the bluer it looks.

Lung

Lungs are our breathing organs. Air breathed out contains more carbon dioxide and less oxygen than air breathed in. When we breathe in, our lungs expand and take in air. When we breathe out the lungs contract and force air out.

Healthy lungs are filled with millions of tiny air sacs called alveoli. Air reaches the alveoli through tiny tubes called bronchioles. In the alveoli the air makes contact with the blood which absorbs oxygen into the bloodstream. Each sac is surrounded by tiny blood vessels (capillaries). The oxygen passes through the thin capillary walls. Carbon dioxide passes out from the blood in the capillaries into the alveoli. It is then breathed out from the lungs.

As the chest gets larger, air is sucked down the windpipe and into the lungs.

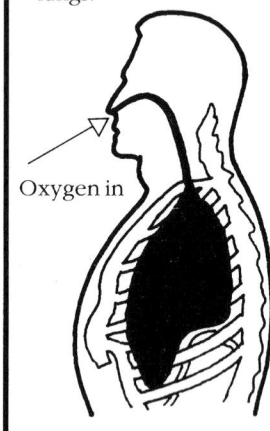

Oxygen in

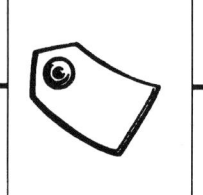

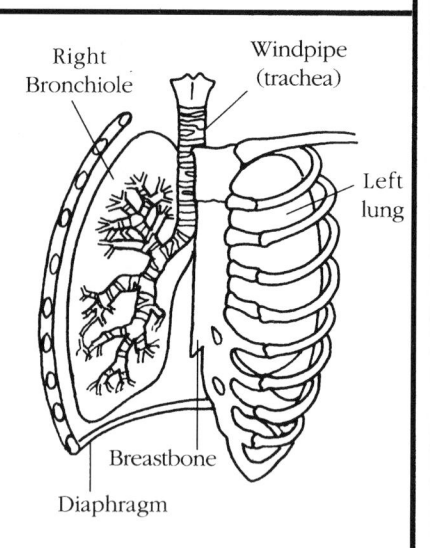

Right Bronchiole — Windpipe (trachea) — Left lung — Breastbone — Diaphragm

Lymph

Lymph is the liquid that carries waste away from our body tissues. Blood carries food and oxygen around the body. When it reaches the tiny blood vessels called **capillaries**, a special liquid called tissue fluid leaks through the very thin capillary walls. This fluid then carries the oxygen and food to the body cells. It also carries away the waste – germs and dead cells – to special capillary tubes called the **lymph capillaries**.

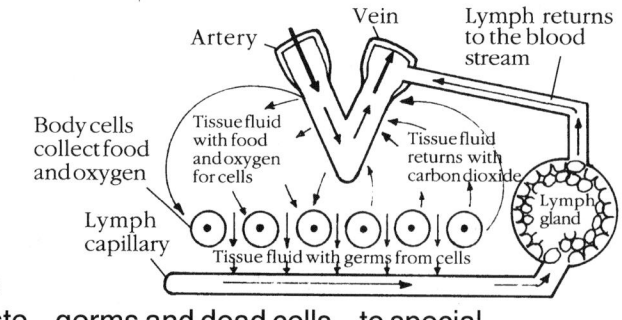

Lymph capillaries are like tiny streams which join to take the lymph into the larger tubes. Eventually these drain into **lymph glands**. Here the germs and dead cells are eaten up by the white blood cells. Clean lymph is returned to the bloodstream through tubes that join a vein in the neck. The whole system is called the **lymphatic system**.

Sometimes people have swollen lymph glands – small lumps in the neck – because their body cannot get rid of the germs and waste quickly enough.

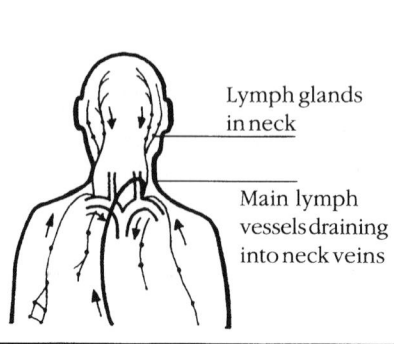

Lymph glands in neck

Main lymph vessels draining into neck veins

Artery — Vein — Lymph returns to the blood stream — Body cells collect food and oxygen — Tissue fluid with food and oxygen for cells — Tissue fluid returns with carbon dioxide — Lymph gland — Lymph capillary — Tissue fluid with germs from cells

Magma

Magma is the molten rock inside the Earth that escapes through a volcano when it erupts. After it reaches the Earth's surface it becomes **lava** – this is the material that flows down the side of a volcano.

As the magma and then the lava cools down it forms different types of **igneous rock**. The type of rock depends on how quickly the molten rock cools down and turns into a solid. **Granite** and **basalt** are igneous rocks.

Magma is mostly liquid but it may contain some solid particles and gas bubbles.

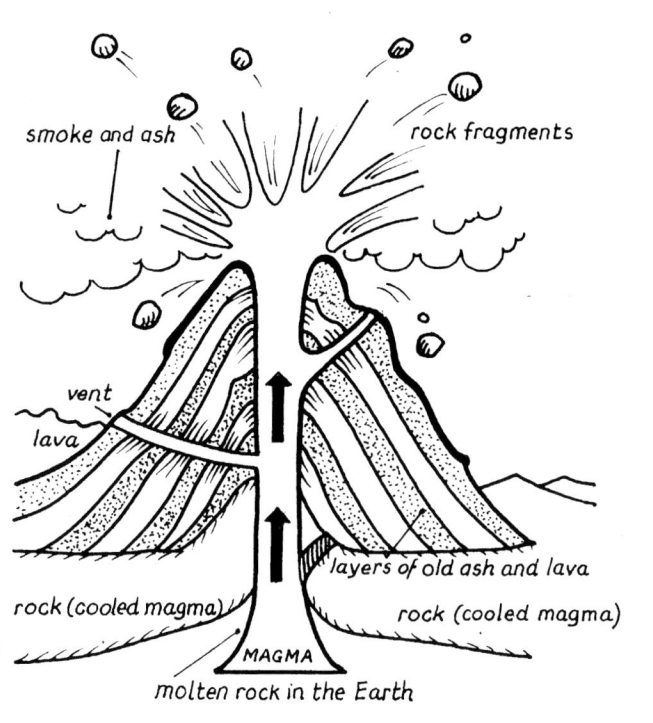

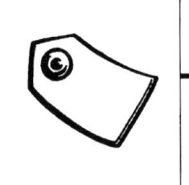

Magnet

A magnet is a piece of material that attracts other types of material, such as iron or steel. Magnets can also attract and repel each other. A magnet in the shape of a bar (a bar magnet) has two ends called the **North pole** and the **South pole**. A North pole attracts a South pole – but two North poles or two South poles push each other apart – they **repel**.

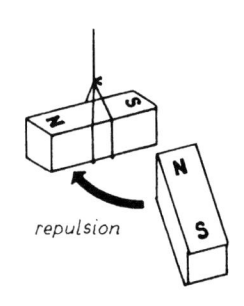

Some magnets (like bar magnets) are **permanent** magnets. They keep their magnetism unless they are heated very strongly or hit for a long time with a hammer. Other magnets are **temporary**. For example, an electromagnet can be made by wrapping a coil of wire round an iron bar – but it is only a magnet when an electric current goes through the wire. Electromagnets can be very useful because their magnetism can be switched on and off.

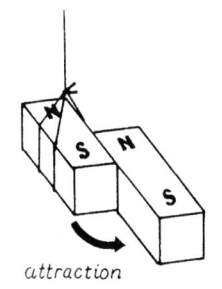

The Earth is magnetic. All magnets have a region of space around them where their magnetism acts – this is called their **magnetic field**. The Earth's magnetic field acts on objects near the Earth's surface – this is how a compass works. Its North seeking pole swings into a particular direction because of the Earth's magnetic field.

Mammal

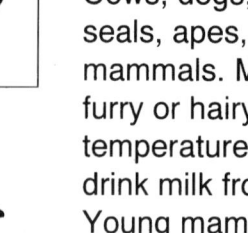

Cows, dogs, cats, whales, dolphins, seals, apes, and humans are all mammals. Mammals have a slightly furry or hairy skin, a constant warm temperature and young mammals drink milk from their mothers' breasts. Young mammals are born alive, except for the platypus which is an egg-laying mammal.

Apes and humans form the biggest group of mammals: **primates**.

Another group is the **marsupials** (e.g. kangaroos, koalas) – these carry their young in pouches.

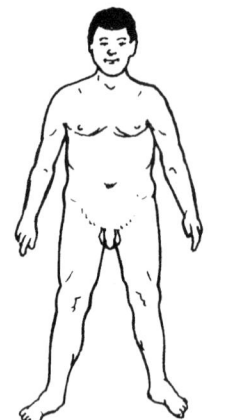

Mass

Mass is the amount of matter or material in an object. It is not the same as **weight**, although the two are linked. The more mass a body has, the larger its weight – weight is the pull of gravity on a body. Weight can change, but mass cannot. If someone has a weight of 600 newtons on Earth, they will only weigh about 100 newtons on the moon – but their mass (the amount of matter in the body) will not change.

Mass is measured in **kilograms** or **grams** (1kg = 1000 grams). A teenager might have a mass of about 50kg and a weight of about 500 newtons on Earth. A heavy person might have a mass of 95kg and a weight of 950 newtons on Earth (about 160 newtons on the moon).

EARTH
mass on Earth = 60 kg
weight = 600 newtons

MOON
mass on Moon = 60 kg
weight = 100 newtons

Melt

Something melts when it turns from being a solid to a liquid. When ice melts it changes from **solid** to **liquid**. The temperature when things normally melt is called their **melting point**.

solid water (ice)

liquid water

SOLID

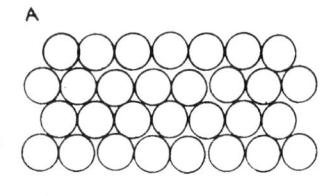

A. Atoms in a solid are arranged in a regular way.

LIQUID

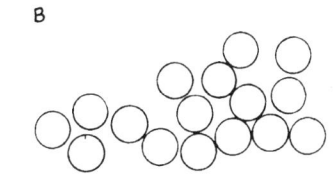

B. This regular arrangement breaks down if the solid is heated till it melts.

Different materials melt at different temperatures. Ice melts (and water freezes) normally at 0 degrees Celsius. Adding salt to ice makes its melting point (and its freezing point) much lower, as low as -10°C. This is why salt is put on the roads in winter, so that the water does not turn to solid ice. Solids, such as steel, lead or iron, all melt eventually when they are heated. Their melting points are much higher than water's.

Inside a solid, the particles are held together in a definite shape - as the solid melts its particles move faster and begin to break away and form a much more fluid shape.

Menstruation

After a girl has become sexually mature (at any age between eight and fifteen years), her body goes through a sequence of events every month called the **menstrual cycle**. The cycle takes about 28 days. In the first five days the womb (**uterus**) loses its inner lining – this passes out of her body with a certain amount of blood. This is called menstruation or more commonly a **period**.

About 13 to 15 days into the cycle, one ovary releases an egg or ovum – this travels slowly along the fallopian tube to the **womb**. Next, the womb grows a new lining which will be used to feed the ovum in case it is fertilised and grows into a baby. If the egg is not fertilised, the lining of the womb breaks down again on about the 28th day and the menstrual cycle begins again. The whole cycle is controlled by **hormones**.

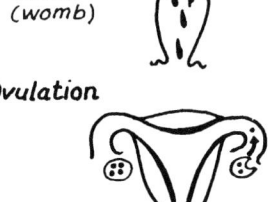

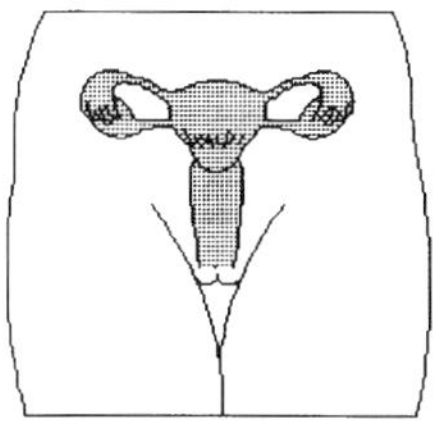

The position of the womb in a female's body.

A. **Menstruation** (womb lining is shed)

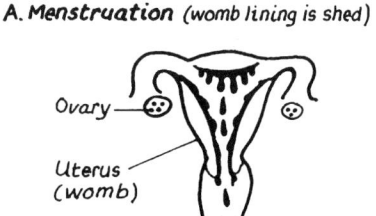

Ovary

Uterus (womb)

B. **Ovulation**

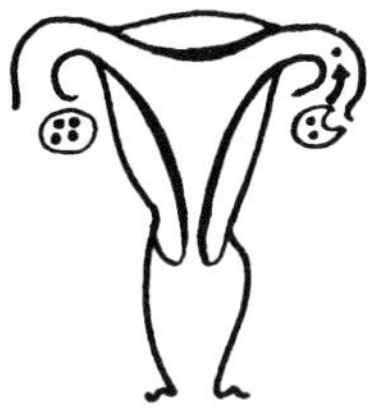

C. **Womb grows new lining**

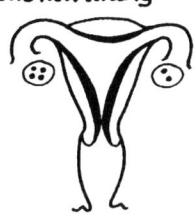

Metal

Copper, tin, gold, aluminium, lead, iron, zinc, silver and magnesium are all metals. Metals are special types of material. Most are good conductors of heat and electricity. They are all solid and hard, except for mercury which is a liquid at ordinary temperatures. Metals can be bent and hammered into shapes and often pulled into the shape of a wire, which makes them very useful. Many metals are shiny but in the air some of them begin to get less shiny or tarnish. Iron does this when it rusts. Metals such as silver and gold don't tarnish and so make good jewellery.

Very few metals are found as pure metal. Most are dug up from the ground as metal ores that look like rocks, for example iron ore (pyrite), lead ore (galena) or limestone (containing calcium). The metal needs to be separated from its ore to make pure metal. Only gold is always found in its metal form.

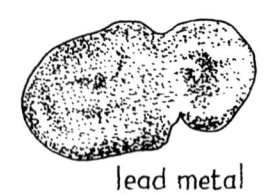

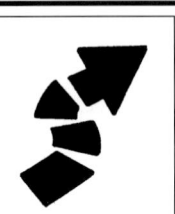

Metamorphosis

This means a complete change (from the Greek for 'change of shape'). Some animals change completely from their baby form when they become an adult. For example, a butterfly egg hatches into a caterpillar (the **larva**) which eats and grows very quickly. The larva develops a hard case (the pupa) with a store of food inside it. The insect develops inside this case and hatches into a butterfly.

A frog also undergoes metamorphosis. Frog spawn develops into tadpoles (the larvae) which change shape completely to become frogs.

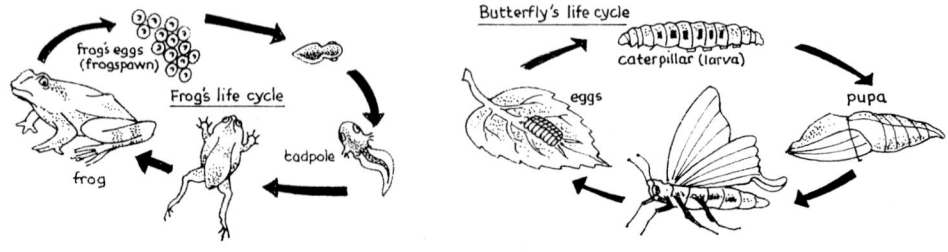

Rocks can also change completely when they are squeezed (compressed) or heated. Limestone can be changed to marble if gets close to very hot **igneous rocks**. Other rocks formed from mud and clay can change completely to become slate. These are called **metamorphic rocks**.

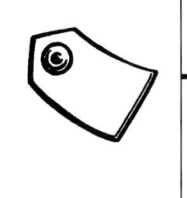

Microbe

The word microbe can mean virus, bacterium, protozoa or some types of fungus. Microbes that cause diseases are commonly known as **germs**.

Viruses are very small microbes – 6 million of them in a row would only be about one mm long. They need a body (a host) to live in where they breed and cause illness. Outside the host they are like dead chemicals but can come to life again. **Bacteria** are larger, as big as five thousandths of a millimetre. They live in many places: the human gut, the sea, in the air.

Some microbes are very useful, such as for making sewage harmless or for ripening cheese. Others are nasty and can cause food poisoning or harmful diseases.

Protozoa are organisms with just one cell – they can cause diseases like malaria or dysentery (very upset stomach). One **fungus** that attack humans causes athlete's foot.

Influenza virus
magnified 10,000 times

Chicken pox virus
magnified 10,000 times

Typhoid bacteria 
magnified 500 times

Pneumonia bacteria 
magnified 500 times

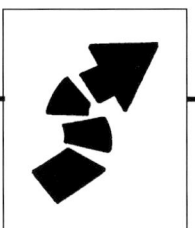

Microwave

Microwaves are short-wavelength (a few centimetres or less) radio waves, in the family of waves called the **electromagnetic spectrum**.

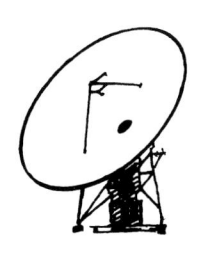

They travel at the speed of light and behave like other waves in the family – they can be reflected and refracted (bent).

Microwaves are used in microwave cookers to heat food – they do this by making the molecules in the food vibrate faster and become 'excited'. But they are also used for communications. Microwave towers transmit information on microwave beams which are received by satellites and transmitted back to different places on Earth.

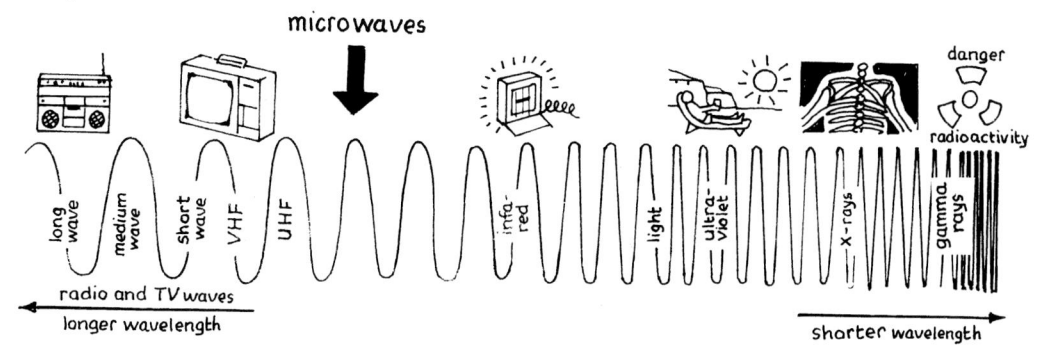

microwaves

long wave | medium wave | short wave | VHF | UHF | infra-red | light | ultra-violet | X-rays | gamma rays

radio and TV waves
longer wavelength ← → shorter wavelength

danger
radioactivity

Mineral

A mineral is a substance that occurs naturally. A lot of minerals are extracted from the Earth's crust by mining, such as rock salt. **Ores** are naturally-occurring minerals – metals such as iron or aluminium can be extracted from them.

Minerals are very important for the health and growth of the human body. Calcium is important for bones and teeth; iron is important for the blood; fluorine for preventing tooth decay. All these minerals are needed in very small quantities in our diet. Most of them can come naturally from fresh fruit, vegetables, milk and cheese.

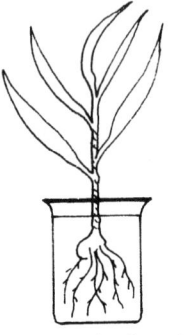

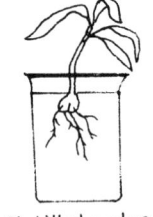

name of mineral	Hardness on Mohs' scale
Diamond	10
Corundum	9
Topaz	8
Quartz	7
Feldspar	6
Apatite	5
Fluorite	4
Calcite	3
Gypsum	2
Talc	1

Plants also need minerals (from the soil) for healthy growth. Nitrogen, potassium, sulphur and phosphorus are all important for healthy crops and plants. These can be provided by **fertilisers** – some are **inorganic** (they don't contain carbon) and artificial. Some fertilisers, like animal dung, are **organic** and, in time, will provide the same minerals for crops.

Water and minerals

Distilled water (no minerals)

Moment

The turning effect of a force is called its **moment**. The moment depends on two things – how big the force is and how far away the force is from the **pivot**. This spanner has a large moment because it is long and the force applied to it is at some distance from the pivot.

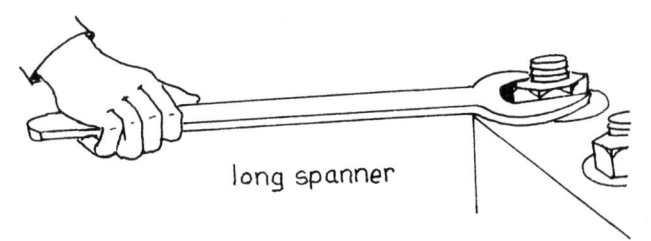

long spanner

Two people on a see-saw, one large and one small, can balance each other if the moments on each side are the same.

The large person has a moment of 800 newtons x one metre, i.e. 800 x 1 or 800 **newton-metres**. The small person has a moment of 400 newtons x 2 metres, i.e. 400 x 2 or 800 newton-metres.

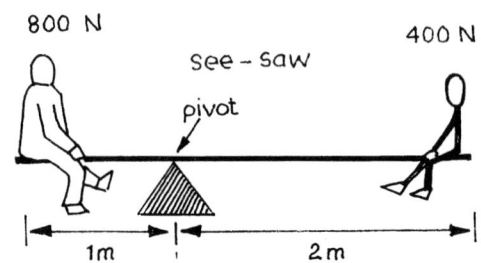

800 N   see-saw   400 N
pivot
1m   2m

A lever such as a crow bar is designed to have a large turning effect or moment.

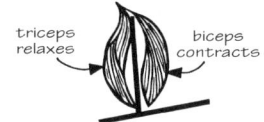

Muscle

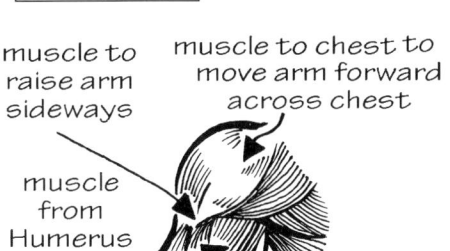

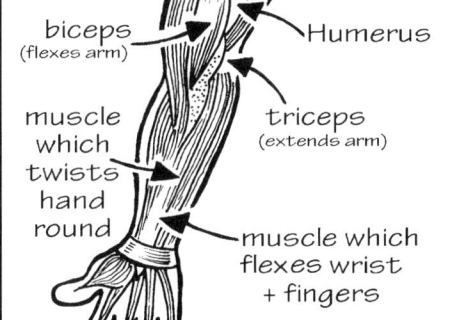

Your body moves because forces act on the bones – these forces come from muscles, over 600 of them in the human body. Muscles contain special types of body cell which can only contract or relax (not lengthen).

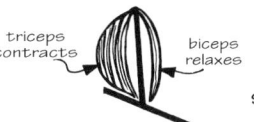

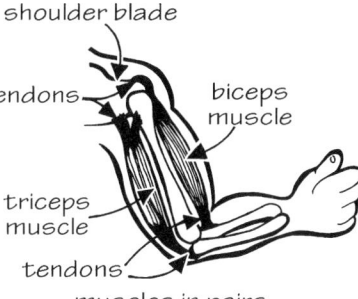

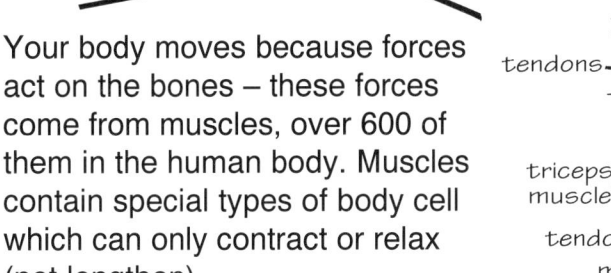

All muscles work in pairs. To raise and bend your arm the **biceps** muscle contracts – but it cannot lengthen itself. Another muscle, the **triceps**, contracts to pull your arm straight again.

Your leg also has a pair of muscles to bend it and straighten it.

Muscles are joined to bones by tendons – these are made up of strong fibres which hardly stretch at all.

Newton

Forces are measured in **newtons**, N for short. **Weight** (the pull of gravity on an object) is a force so it is also measured in newtons. Here are the sizes of some typical forces, given roughly in newtons:

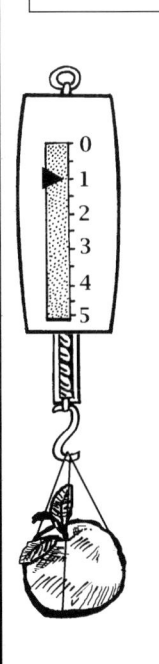

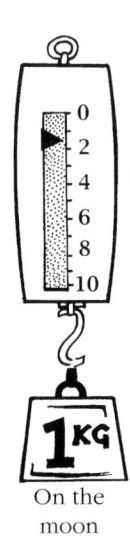

weight of an average apple ..... one newton (1N)
pulling force needed to
open a drinks can .... 20 N
force to open a heavy door ..... 30 to 40 N
pulling force of
a tug-of-war team............. about 5000N
weight of a large man .... over 900 N
force from a jet engine........ over 150,000 N

Forces are measured with **a newton meter** (it looks like a spring balance).

A newton meter can be used to measure weight. A mass of 1kg weighs about 10 newtons on Earth, but less than 2 newtons on the Moon.

Nitrogen

Nitrogen is a colourless gas with no smell which makes up about 4/5 of the air we breathe.

It is vital to all living things because it is an essential part of protein. The Earth's nitrogen goes round in a continuous cycle, from plants to animals, into the air, back to the soil and into plants again. This is called the **nitrogen cycle**.

Some bacteria take nitrogen out of the soils and into the air – other bacteria put it back in again (nitrogen-fixing bacteria). Some plants, such as peas, beans and clover, have special 'nodules' on their roots which let them use nitrogen directly from the air.

Other plants take in nitrogen through their roots from nitrates in the soil. Animals eat plants and when the animals die or excrete (go to the toilet) some of the nitrogen goes back to the soil as the remains decay. The whole cycle goes on all the time and keeps plants and animals alive and healthy.

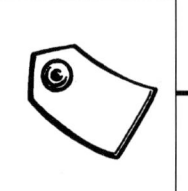

ROOT NODULES OF A BEAN PLANT WHICH "FIX" NITROGEN

Nucleus

The very centre of an atom is called its **nucleus**. It occupies a very small space compared to the rest of the atom.

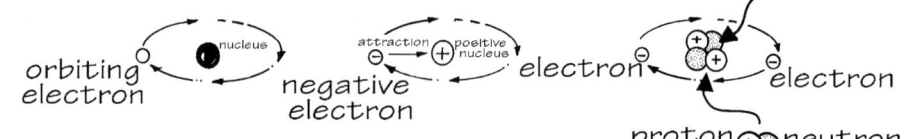

The nucleus is made up of two tiny particles, the **proton** and the **neutron**. Protons have a positive charge which balances the negative charge on an **electron**. Neutrons are neutral (hence the name) but they do add to the mass of an atom. The total number of protons and neutrons in the nucleus is called the **mass number** of an atom. For example, a very large atom like Uranium has a mass number of 238 (92 protons and 146 neutrons).

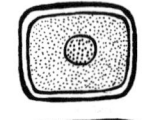

The nucleus of large atoms like uranium can be split to release nuclear energy – this is called **nuclear fission**.

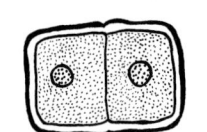

Animal and plant cells also have a nucleus.

The nucleus of a cell is its 'control centre'. The nucleus of a cell can divide to make two new cells.

CELL DIVISION

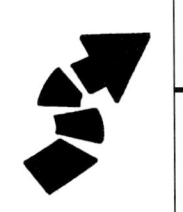

Nutrition (nutrient)

Nutrition means food or feeding, either by plants or animals. All living things need food to give them energy, to keep their body cells working and to build or repair living tissues.

Humans need a range of nutrients in their food: **carbohydrates**, **proteins** and **fats**. Carbohydrates and fats give us energy, proteins are needed for growth and to repair the body tissue. We also need tiny amounts of **minerals** and **vitamins** – and, of course, lots of water to live.

Animals eat their food and break it down by digestion – but plants build their own nutrients from the carbon dioxide in the air, with the help of sunlight (photosynthesis). The carbon in a tree or plant comes from the carbon dioxide in the air – only the minerals and water they need come from the soil. Fungi and parasites feed in slightly different ways.

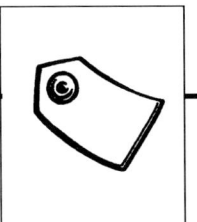

Ohm

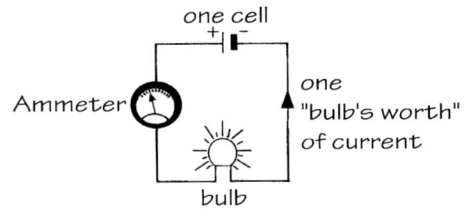

An ohm is the unit used to measure the **resistance** in an electric circuit. The bigger the resistance in a circuit, the more voltage needed to make a current flow in it. One bulb in a circuit may be lit up by one cell.

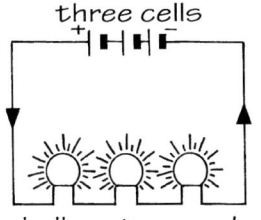

But if two bulbs are put in side-by-side (in series), the resistance is doubled. Now two cells are needed to make the same current flow. Three bulbs will have three times the resistance and will need three cells to light them up.

Each bulb added makes more resistance in a circuit like this. More and more resistance will make the current weaker and weaker, unless we increase the voltage.

Resistance, voltage and current are connected by a simple equation:

$$\text{resistance (in ohms)} = \frac{\text{voltage (in volts)}}{\text{current (in amps)}}$$

Very thin wires have much more resistance than broad, thick wires.

Ohm symbol Ω (Omega)

Omnivore

An **omnivore** is an animal that eats both plants and animals. Human beings are omnivores, as are some birds and bears.

Some animals, for example rabbits or naturally fed cows, eat only plants such as grass. They are herbivores. Herbivores are called **primary consumers**.

Other animals are flesh-eaters or carnivores, e.g. foxes. They eat herbivores such as rabbits or hens and are called **secondary consumers**. They may also eat small carnivores and are then called **tertiary consumers**.

Omnivores are sometimes primary consumers, e.g. when we eat plants, and sometimes secondary or even tertiary consumers.

Orbit

An **orbit** is the curved path that a moving object takes. It may be a circular path, like this person swinging a ball and chain in a circle:

An orbit may be in the shape of an ellipse (oval shape). The planets in our solar system travel around the sun in orbits that are ellipse shaped:

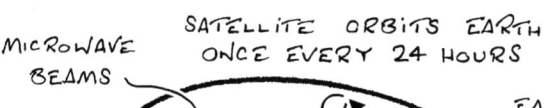

Comets like Hale-Bopp also travel in curved orbits. The Moon orbits the Earth in 28 days – the Earth orbits the sun in about 365 days. Artificial satellites used for communications and other things orbit the Earth. Some are in special orbits so that they are always over the same spot on Earth – these are called **geo-stationary orbits**.

In the atom, electrons orbit the nucleus rather like the Earth round the Sun:

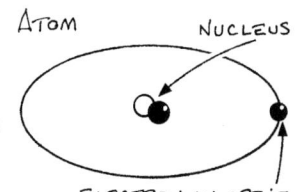

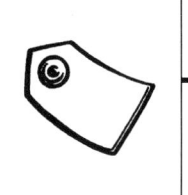

Organ

An **organ** is a part of a plant or an animal that does certain important jobs. In a plant, the leaves and the roots are organs:

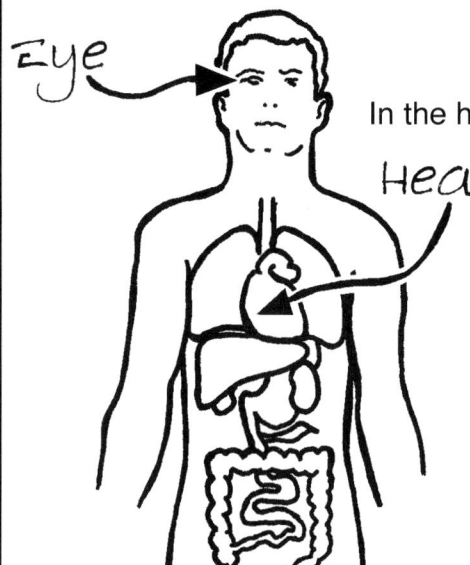

In the human body, there are lots of organs such as: the heart, liver, kidney, pancreas, stomach, eye and so on.

Organs often work together in **organ systems**. For example, the **digestive system** in a human has the stomach, liver, pancreas and other organs all working together as a group. The **circulatory system** is made up of the heart, the blood and the blood vessels that carry blood.

Some organs have been successfully moved from one animal to another – this is called an **organ transplant**.

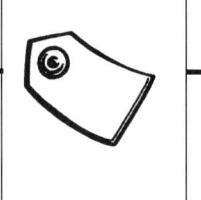

Organism

An **organism** is a living thing. Organisms can be divided up or classified into various types:

All organisms have seven things in common, which they all do (their seven 'vital functions'): **grow**, **excrete**, **respire**, **move**, **respond**, **feed** and **reproduce**. An eighth thing in common is that they are all made up of cells.

Ovary

An **ovary** is the female organ in a plant or an animal that makes eggs. In a flowering plant the ovary makes **ovules**. Ovules eventually become seeds to make new plants:

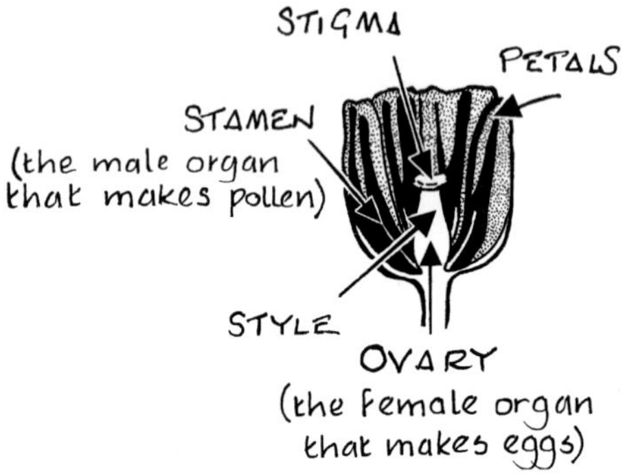

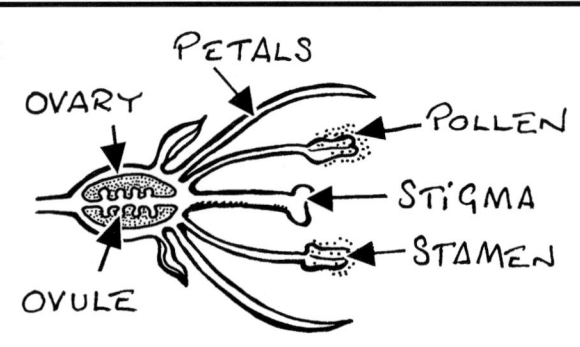

In a mature female, the ovary makes the ovum or egg, which leaves the ovary once each month and travels down the **fallopian tube** to the womb or uterus.

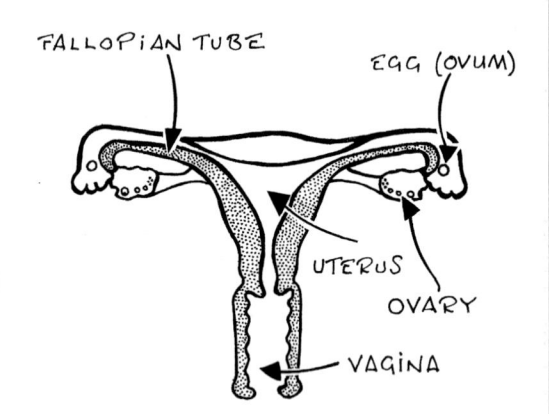

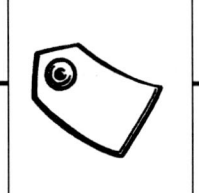

Oxygen

Oxygen is the invisible gas in the atmosphere which keeps us, and all other plants and animals, alive. It makes up about one-fifth of the air we breathe.

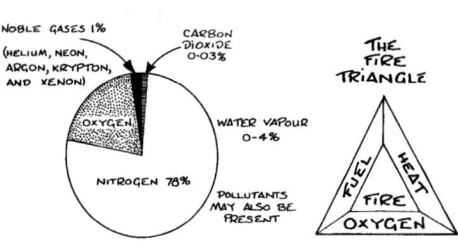

It is needed in our blood (and other animals' blood) for **respiration**. This gives us the energy to stay alive.

Glucose + Oxygen > Carbon Dioxide + Water + Energy

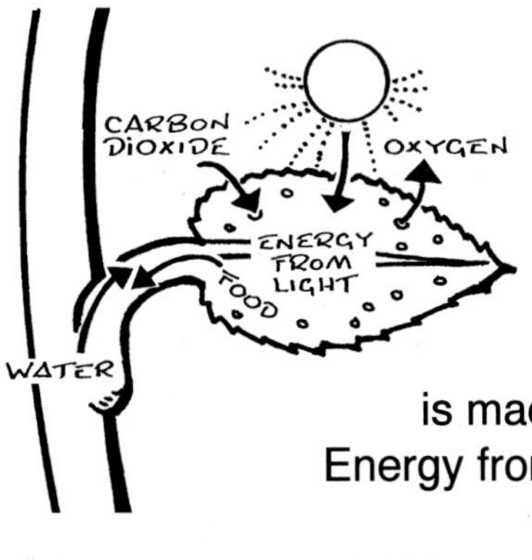

Oxygen is also needed for burning or **combustion**. Fuel burns in air to make energy, just like respiration in human cells.

Fuel + Oxygen > Heat Energy + Waste Products

Many fuels, like coal and gas, burn in oxygen to give energy. Oxygen is used up by respiration and burning, and carbon dioxide is made. Oxygen is put back into the atmosphere by **photosynthesis**. Energy from sunlight uses carbon dioxide from the air to make oxygen.

Ozone

Most oxygen atoms join up in pairs to make the oxygen molecules we breathe, O_2. Ozone is a special form of oxygen with molecules made up of triplets of oxygen atoms, O_3. Ozone is a very important gas in the higher parts of the atmosphere.

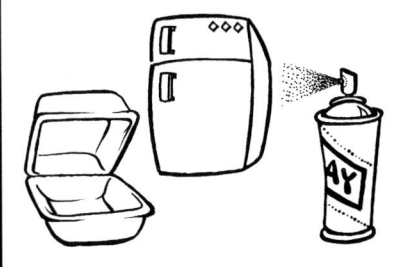

About 20 to 50 km above the ground, it forms a special protective layer called the **ozone layer**. This protects animals and plants by 'filtering out' the harmful, high energy **ultraviolet rays** that come from the sun. These rays can harm plants and animals on Earth if they get through. For example, high energy ultraviolet rays can cause skin cancer. They can also affect crops like rice and wheat and make them less healthy.

The ozone layer has been damaged by some chemicals, such as **CFC**s. These come from things like aerosol sprays.

When CFCs rise up into the stratosphere they reduce ozone (O_3) to ordinary oxygen (O_2). This does not work as a filter for harmful rays - so more of them get through. Eventually, there is sometimes a 'hole' in the ozone layer over certain parts of the Earth (especially the South Pole).

Parallel

A parallel beam of light is produced by a strong torch or a searchlight. The searchlight needs a specially-curved mirror to make the beam parallel, not spreading out or diverging:

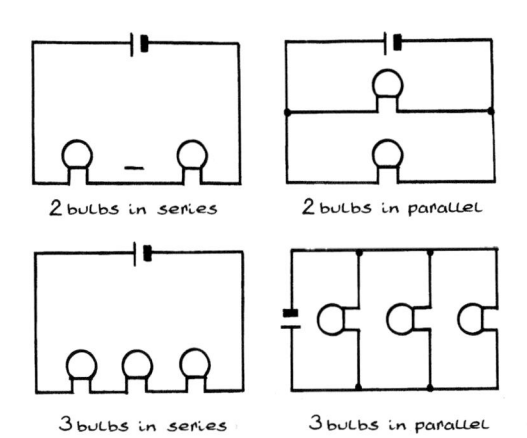

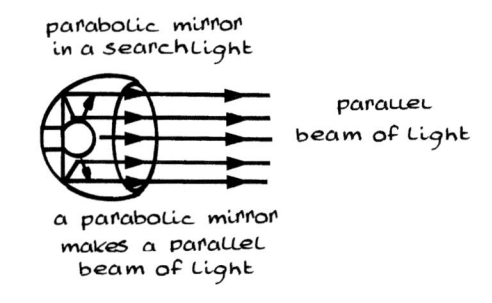

parabolic mirror in a searchlight

parallel beam of light

a parabolic mirror makes a parallel beam of light

2 bulbs in series

2 bulbs in parallel

3 bulbs in series

3 bulbs in parallel

In an electric circuit, light bulbs can be connected in **series** or in **parallel**. In series, if one bulb is taken out, the other does not light because the circuit is broken. But if bulbs are connected in parallel, and one is taken out, there is still a complete circuit for the electric current to go through the other bulb, and it stays on.

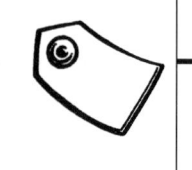

Parasite

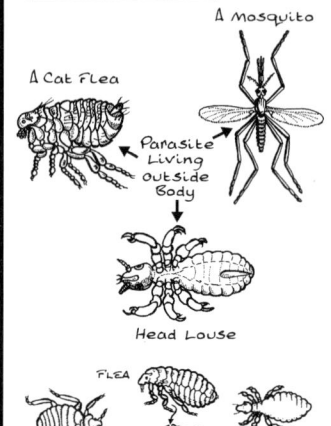

A parasite is an organism that lives on or in the body of another living organism. The organism on which the parasite lives is called the **host**. The host gets no benefit from the parasite and is sometimes harmed by it.

Fleas, mosquitoes and lice are parasites that live on the outside of the host, such as a cat, dog or human. They get their food by piercing the skin with a hollow tube on their body (like a needle) and sucking blood out. Leeches and hookworms also suck blood but they use teeth which break the skin and make it bleed.

Some parasites live inside the host. Germs like viruses or protozoa are parasites and can cause serious illnesses. Tapeworms live inside the human intestine and eat some of the digested food. Liver flukes live inside animals like sheep or cows and suck their blood.

Plants have parasites too: Greenfly live on leaves; Mistletoe plants live on trees.

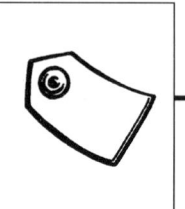

Particles

Everything is made up of tiny bits called particles.
Particles are too small to be seen with a microscope, but scientists believe they make up every substance. This is called the **Particle theory of matter**.

Particles are always moving – the hotter something gets, the faster its particles move. (They stop moving completely at a temperature called absolute zero, – 273°C, which has never been reached).

Particles behave differently in solids, liquids and gases.

Particles in a solid attract each other quite strongly and are not as free to move. In a liquid, their attraction is weaker and they have more freedom. In a gas, particles hardly attract each other at all and can move anywhere inside their container.

Particles are either atoms or molecules.

Periodic Table

The periodic table is a way of grouping or classifying all the elements that we know of in an orderly way. It is made up of eight columns and seven rows. Each element has its own special symbol, for example Zn is zinc.

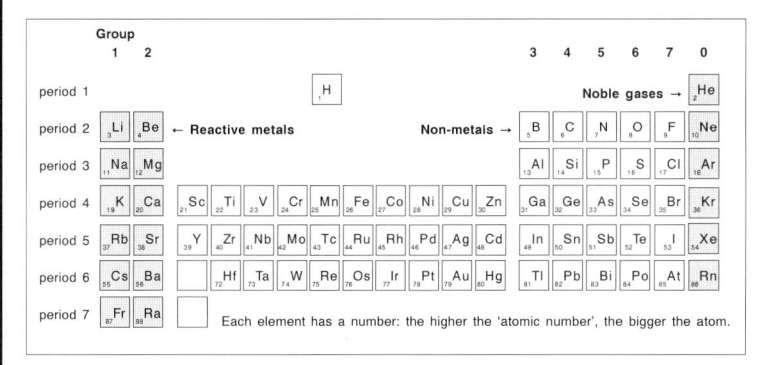

Each column is called a **group** and these run from group 1 to group 7 with group 0 on the far right. All the elements in a group are similar. For example, group 1 is made up of metals that all react with water. Group 0 is made up of the 'noble gases': helium, neon, argon and so on. The rows are called **periods**. As you move down the group to different periods, the way each element behaves changes slightly. For example, the metals in group 1 react much more violently with water as you go down to higher periods. You can tell what any element is like and how it behaves from its position in the table.

Each element has a number, its **atomic number**. The higher the number, the bigger the atom, and the more protons in its nucleus. For example, copper (Cu) is number 29 and has 29 protons in its nucleus.

Photosynthesis

Photosynthesis is the process by which plants change light energy into stored, chemical energy. The green material in leaves – **chlorophyll** – absorbs energy from sunlight and changes water and carbon dioxide into glucose and oxygen.

Photosynthesis is vital to life on Earth. It takes **carbon dioxide** out of the air and puts **oxygen back**. It provides plant food for herbivores, which are the food of **carnivores** (all of which support parasites). Photosynthesis is the beginning of all food chains.

The Earth's forests are vital for the photosynthesis that their trees provide. Most of the world's oxygen comes from them and they also help to absorb some of the carbon dioxide produced from burning fossil fuels.

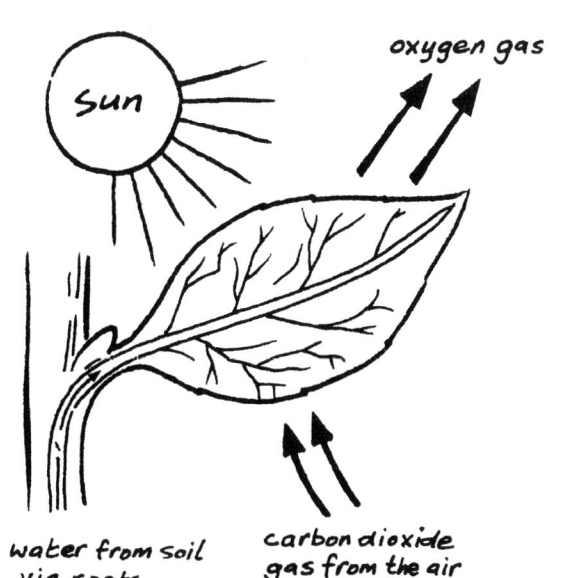

Pivot

A lever turns around a point called a pivot. A long lever resting on a pivot can be used to lift a heavy weight.

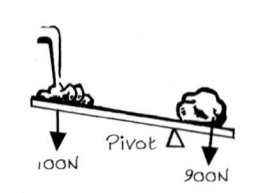

A small force can be used to lift a big weight because it is a long way from the pivot or fulcrum. With a long spanner the force is a long way from the pivot and this gives it a greater turning effect.

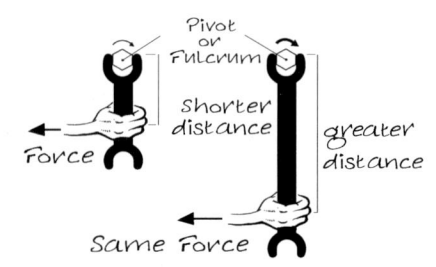

The turning effect is called the **moment** of the force. It depends on two things: the **size** of the force and its **distance** from the pivot.

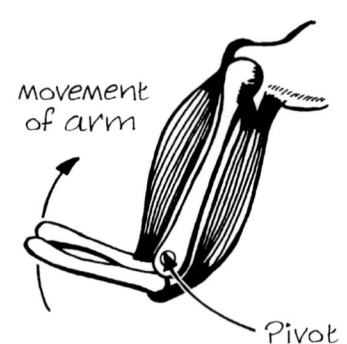

The joints in the human body, such as the knee or elbow, act as pivots.

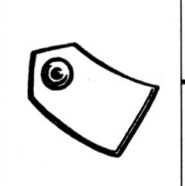

Placenta

The placenta is the organ in a mother's body which supplies a developing baby in her womb with food and oxygen, and removes the waste products. It is a disc-shaped organ attached to the wall of the womb. The developing baby (the foetus) is joined to the placenta by the **umbilical cord**.

The baby's tiny heart pumps blood through the umbilical cord into the placenta. In the placenta, the baby's blood absorbs food and oxygen from its mother's blood; and releases carbon dioxide and other waste products into the mother's blood. Her blood system carries this away to be excreted. So, the placenta takes the place of the baby's own lungs, kidneys and digestive system.

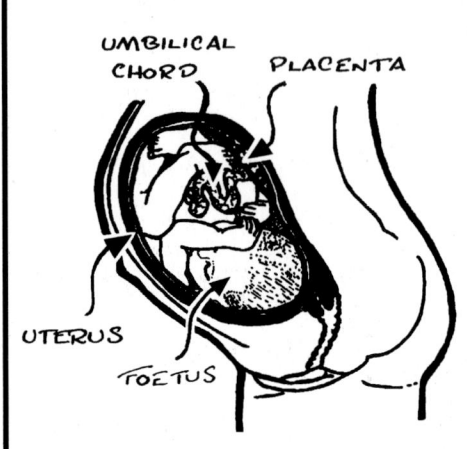

After the baby is born the umbilical cord has to be cut (this is where your 'belly button' came from). Most mammals have placentas for their young – wild mammals bite through the cord after birth. The placenta is pushed out of the womb after the baby has been born – this is called 'after-birth'.

Planet

A planet is a large object that goes round and round the Sun in orbit. The Earth is a planet. Planets are kept in their orbits by the Sun's gravity. There are nine planets orbiting the Sun, and together these make up the Solar System.

The order of the planets is:

MERCURY, VENUS, EARTH, MARS (the inner planets) . . . JUPITER, SATURN, URANUS, NEPTUNE, PLUTO.

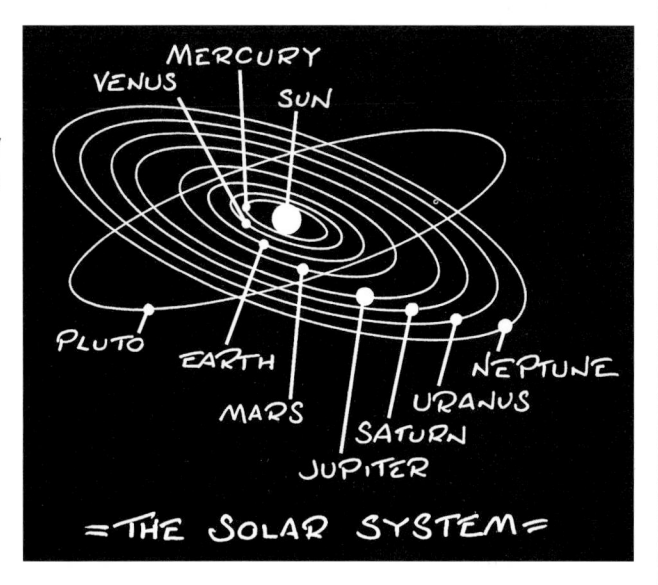

A simple rhyme for remembering this order is:
" May Vickers Eats Mark's Jam Sandwiches Until Nellie's Party"

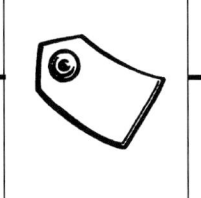

Pollen

Pollen grains contain the male sex cells (or **gametes**) of a flower.
The **stamen** of a flower contains the pollen grains in four sacs.

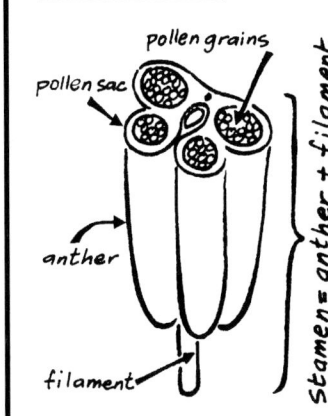

The **stigma** of a flower receives the pollen which then goes down a pollen tube to fertilise the **ovules**.

Pollen is carried from the stamen of one flower to the stigma of another either by wind or by insects such as bees. Wind pollinated flowers such as grass and stinging nettles, have a lot of light pollen which is carried by the air. Insect pollinated plants, such as buttercup and dandelion, have fewer but larger pollen grains and the flowers are very colourful so that they attract insects. Pollen from the stamen rubs off onto the insect as the insect feeds from the nectar of the flower. It comes off again when the insect brushes against the stigma of another flower.

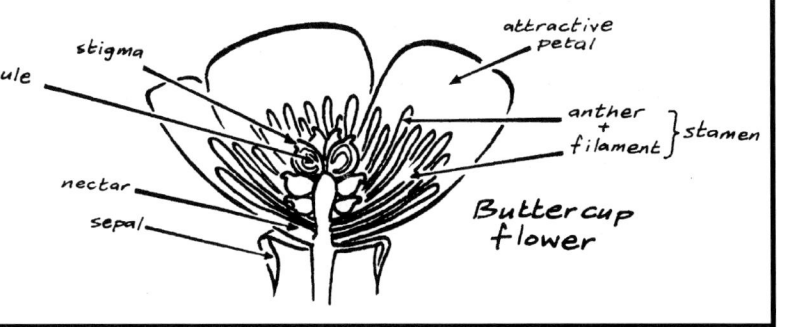

Pollution

Pollution is any kind of damage to the environment we live in. We suffer from air, water, radiation and noise pollution.

Most air pollution comes from burning **fossil fuels**, like coal and oil. Cars use petrol (which comes from oil) and their exhaust gases are poisonous. Leaded petrol is worst because it puts lead into the atmosphere. Using fossil fuels produces **carbon dioxide** and oxides of nitrogen (burning coal also makes sulphur dioxide). These oxides rise into the atmosphere and react with water to make **acid rain**.

Some water pollution comes from human sewage, if it is not properly treated. Some comes from detergents, some from oil spills in the sea. Some pesticides which farmers use on their fields get washed away into the streams and rivers. Some water pollution comes from factory waste.

Power

A person or a machine is powerful if it can do work quickly. The faster they do work, the more powerful they are. If two people are lifting bricks onto a truck, the one who loads more quickly is more powerful.

Power can be calculated by working out how much work is done in each second:

POWER (in watts) = $\dfrac{\text{WORK DONE (in joules)}}{\text{TIME TAKEN TO DO IT (in seconds)}}$

The person who does work more quickly has more power

This crane lifts bricks weighing 4000N up by 2 metres. The work done is 4000N x 2m = 8000 joules. The crane takes 4 seconds, so its power is:
$\dfrac{8000}{4}$ = 2000 watts.

A crane is more powerful than several people

Predator

A predator is an animal which feeds by eating another animal called its **prey**. A fox is a predator, its prey might be a rabbit or a hen. Big fish (predators) eat small fish (prey). Lions and hawks are predators (bird predators are called raptors).

Predators and their prey form part of a **food chain**.

Sometimes this can be drawn like a **pyramid** with the chief predator at the top, like those shown here.

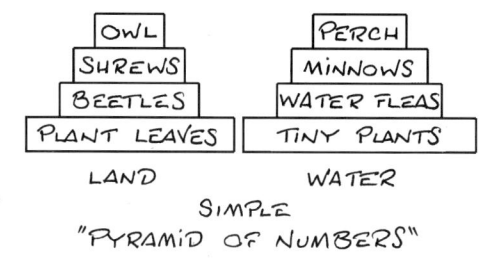

```
OWL              PERCH
SHREWS           MINNOWS
BEETLES          WATER FLEAS
PLANT LEAVES     TINY PLANTS
   LAND             WATER
         SIMPLE
   "PYRAMID OF NUMBERS"
```

Life is not always this simple though – some animals such as insects are preyed on by more than one predator.

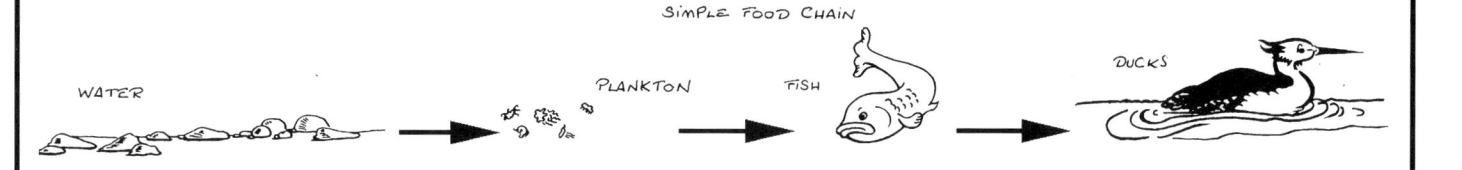

SIMPLE FOOD CHAIN

WATER → PLANKTON → FISH → DUCKS

Pressure

The pressure from a force depends on how large an area it is spread over. If you push a coin on its side into Plasticine it goes further in than if you push on its face. The coin on its side exerts a larger pressure. People wear snowshoes in deep snow so that their weight is spread over a larger area, the pressure is less, and so they don't sink into the snow.

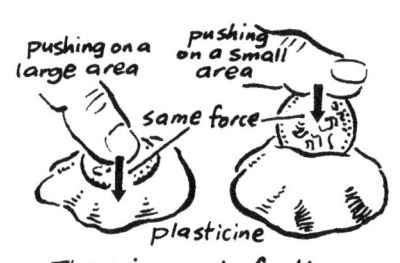

pushing on a large area / pushing on a small area / same force / plasticine

The coin goes in further when the pushing force is concentrated onto a small area.

The air around us has pressure, called **atmospheric pressure**. You can see this when the air is pumped out from inside a can – it collapses from the air pressure outside squashing it. Air pressure is strong enough to support the water in a glass.

Glass full of water / piece of card / Atmospheric pressure

As you go higher and higher in the atmosphere the pressure of the air gets less and less.

Liquids have pressure too. As you go deeper and deeper in a liquid, the pressure gets higher and higher. This is why deep sea divers have to wear special suits, and dams are built much thicker at the bottom.

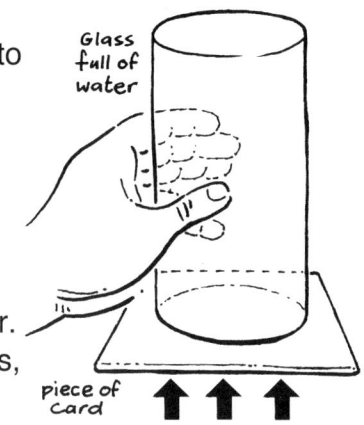

Prism

A prism is a piece of glass in the shape of a 3D triangle which can do interesting things with light. Some prisms can split light into the seven colours which make up white light.

In a rainbow, raindrops act like tiny prisms and split sunlight into the colours of the **spectrum**: RED, ORANGE, YELLOW, GREEN, BLUE, INDIGO, VIOLET (remember: 'Richard of York gave battle in vain')

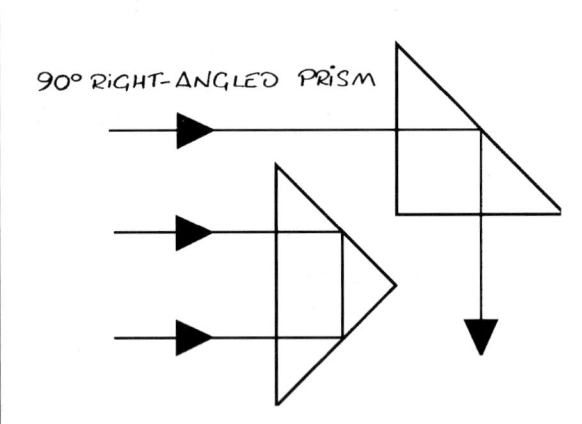

If the seven colours are painted evenly onto a disc and the disc is spun quickly, it appears to be white.

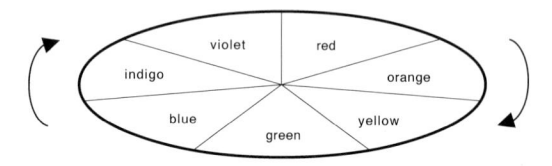

Prisms shaped so that one angle is 90° and the others are 45° can be used to reflect light, or change its direction completely. Prisms like this are called right-angle prisms and are used in cameras and binoculars.

Protein

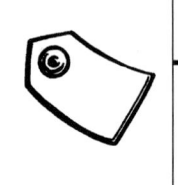

Protein is a body building food. It is used for growth (especially for young people or for pregnant women), or for repairing damaged or worn-out body tissue. Proteins can also be a source of energy. They are in foods like meat, fish, eggs, cheese and Soya beans; and also flour, rice and oatmeal.

Vegetarians can get all of the proteins they need if they eat a wide variety of plant foods such as beans, peas, brown rice, pasta and nuts.

Proteins have very large molecules made up of lots of different amino acids joined together.

When our body digests proteins it breaks them down into amino acids. Like protein, most of the human body is made up of the four elements: carbon, oxygen, hydrogen and nitrogen.

Pulse

A pulse is a tiny ripple of energy. Every time the human heart pumps blood into your main artery, it causes a ripple to move along the walls of the artery. This ripple is your pulse. You can feel it in the artery in your wrist or at another 'pressure point' around your body such as in your neck.

SOME OF THE BODY'S PRESSURE POINTS

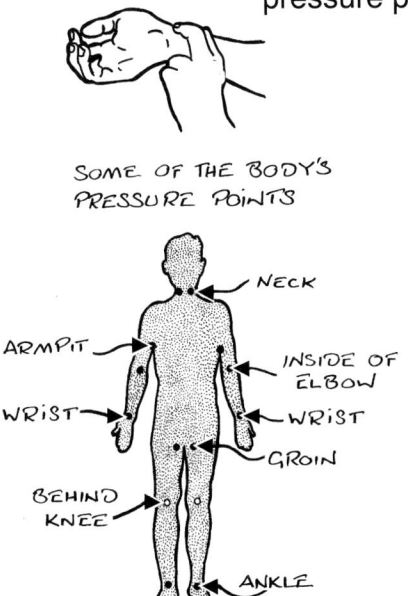

ARMPIT
WRIST
BEHIND KNEE
NECK
INSIDE OF ELBOW
WRIST
GROIN
ANKLE

The human heart beats about 70 times each minute and this is called your pulse rate. A new-born baby's pulse rate can be as high as 140 beats per minute. It can be as low as 50 per minute for some people (if they are very fit) or as high as 150 or even 200 pulses per minute if someone is very excited, frightened or has just done some heavy exercise. This means that a person's heart beats about 100,000 times per day, and as many as 2,500 million pulses in their lifetime. But compare this with a sparrow – its pulse rate is 500 per minute!

Some special stars send out quick, short bleeps of radio waves. These are called **pulsars**, and their pulses of energy can come every fraction of a second. One famous pulsar is at the centre of the Crab Nebula, a huge patch of glowing gas in the sky. It sends out 30 pulses each second.

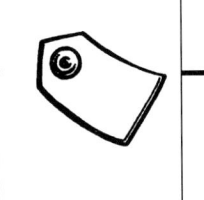

Radiation

Moving rays are called radiation. The most important family of rays is the **electromagnetic spectrum**, made up of radio waves, infrared, light rays, ultra violet, X-rays and gamma rays. They are all types of **electromagnetic radiation**. X-rays can be used to 'see' inside the body because they can travel through solid material such as human flesh, but not bone, so the bone is highlighted.

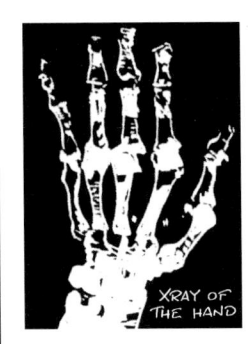

XRAY OF THE HAND

Three types of radiation come from the nucleus of some atoms. These atoms e.g. uranium, plutonium, are called **radioactive**. The three types are called **alpha**, **beta** and **gamma**.

Alpha is actually a tiny moving particle, made up of two protons joined to two neutrons; beta is a moving electron; gamma is a ray from the electromagnetic family. Each type can be harmful to humans, but alpha particles can be easily stopped. Beta can be stopped by aluminium, but gamma rays can only be stopped by very thick lead or concrete:

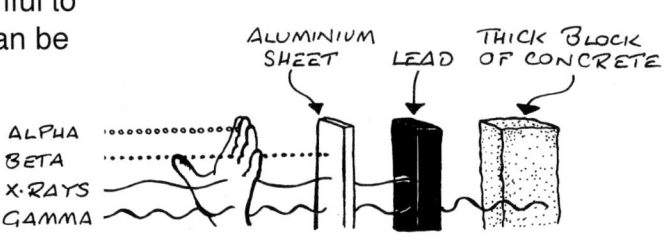

ALUMINIUM SHEET
LEAD
THICK BLOCK OF CONCRETE

ALPHA
BETA
X-RAYS
GAMMA

Reaction

If two substances react together they make a completely new substance. This is called a **chemical reaction**. Sulphur can react with oxygen to make a new compound called sulphur dioxide. Hydrogen reacts with oxygen to make water.

There are hundreds of chemical reactions. If coal burns in air, the carbon in it reacts with oxygen to make carbon dioxide. This is a **combustion reaction**. When the iron in a screw rusts in damp air it makes iron oxide – this is a **corrosion reaction**.

The rusting of iron is a slow **chemical reaction**. Iron, water and oxygen from the air are the **reactants**.

Iron (makes up nearly all of steel) + Water + Oxygen → Rust

Rusting is quite a slow reaction that may take days. Some reactions are much quicker. If you drop a metal like potassium or sodium into water it reacts immediately and fizzes around making hydrogen gas. Metals are very reactive.

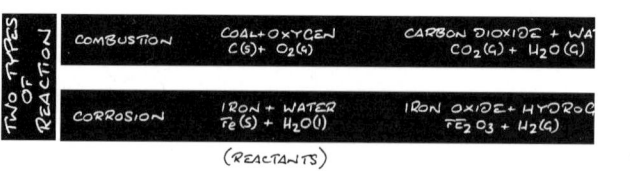

TWO TYPES OF REACTION

COMBUSTION	COAL + OXYGEN $C(s) + O_2(g)$	CARBON DIOXIDE + WATER $CO_2(g) + H_2O(g)$
CORROSION	IRON + WATER $Fe(s) + H_2O(l)$	IRON OXIDE + HYDROGEN $Fe_2O_3 + H_2(g)$

(REACTANTS)

Reactions can often be speeded up. Warming things up can make them react more quickly. Light can help a reaction to speed up, e.g. photosynthesis. Adding a **catalyst** (or enzyme) can speed up some reactions.

Reflection

Rays or waves are reflected when they meet a hard surface and bounce back off it. When sound waves are reflected they make an **echo**. Heat rays (infrared) are reflected by shiny surfaces.

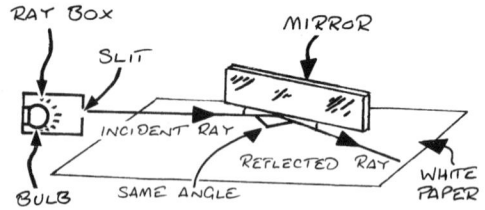

Light rays are also reflected from hard, shiny surfaces such as a mirror. The angle the ray comes in at (the **incident angle**) is the same as the angle it bounces off at (the **reflected angle**).

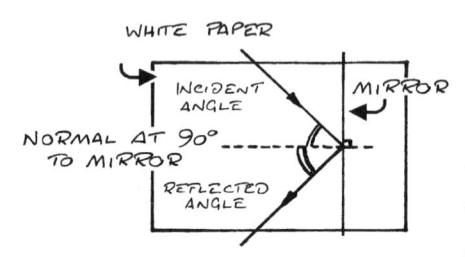

We can see some objects (like a fire, a light bulb or the Sun) because they give out their own light – they are **luminous**. But we see most objects because they reflect light (from a luminous object like the Sun or a light bulb) into our eyes.

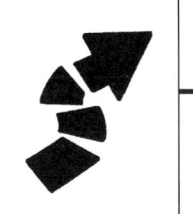

Refraction

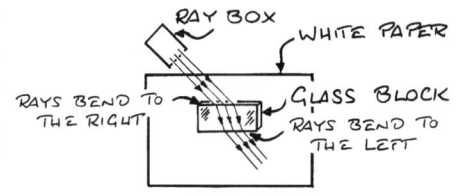

Rays or waves usually travel in a straight line at a steady speed. But if they travel into another substance (a medium) they sometimes bend or change direction. This bending is called **refraction**. When a beam of light goes into a glass block it bends one way – as it leaves the block it bends the other way.

Water bends (refracts) light rays too. This is why a stick looks bent when it is lowered into water. A coin on the bottom of a cup can be made to suddenly appear as you pour water into the cup. The water bends light rays coming from the coin so that they reach your eye.

Glass lenses work because they refract or bend light rays as they go through them. A **convex lens** will focus light rays from the Sun on to one point (the focus) and can burn a piece of paper.

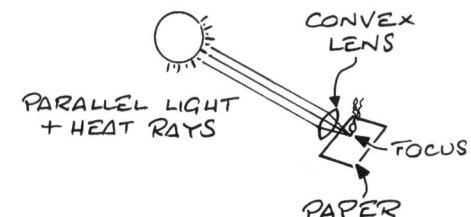

Renewable

Some sources of energy can be used again and again – they are called **renewable sources of energy**. Energy from wind is renewable – it can be used to turn giant wind turbines which generate electricity.

Tides in the sea are renewable sources – they go in and out every day, and this movement can be used to generate electricity.

The energy of the sun (solar energy) is also renewable – it can be collected using solar panels, or focused by using large mirrors to concentrate the sunlight (a solar furnace).

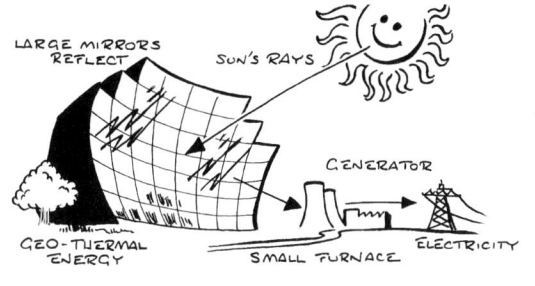

Energy from moving water in rivers is renewable – this is hydroelectric energy. Nuclear energy, energy from the waves of the sea and energy from the gas from rotting material (biogas) are all renewable sources.

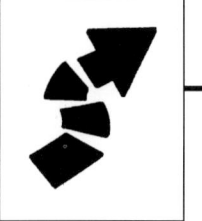

Reproduction

All living things need to make new, young organisms just like themselves – this is called reproduction. If it did not reproduce, that type of living thing would die out (become extinct). **Sexual reproduction** needs a female cell and a male cell. In flowers, the **pollen** from the male part fertilises the **egg** or ovule of the female part, to make seeds which form next year's plants.

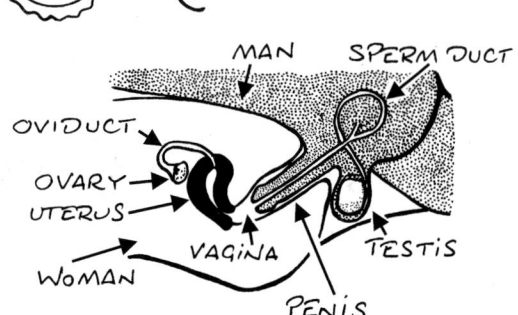

In humans, the **sperm** from a male enter a female's body during sexual intercourse. One of the sperms enters a female egg cell and fuses with it. The egg eventually turns into an **embryo** which becomes a new animal.

In fish, frogs and reptiles the embryo develops in an egg, feeding from the egg yolk. In mammals the baby grows inside the mother's body and feeds from the placenta until birth.

Some living things reproduce without sex – this is called **asexual reproduction**. For example, amoebas just split into two.

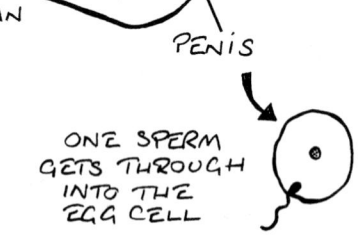

Reptile

Snakes, lizards, tortoises and crocodiles are all reptiles. The dinosaurs were reptiles.

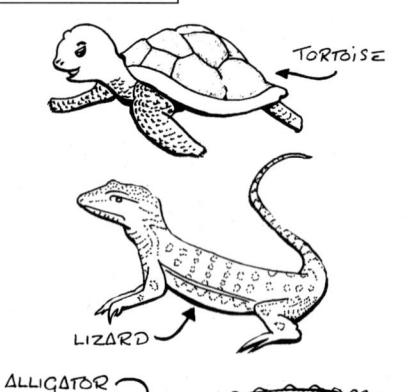

Reptiles :

- have a dry, scaly skin;
- have a backbone
- breathe with lungs
- lay eggs with a tough, leathery shell
- produce young that look like the adult, i.e. not a larva

They have short, strong limbs and can move or 'dart' very quickly, but not for long. Reptiles are cold-blooded – that means that their body temperature can change and they move into hot places to warm up and colder places to cool down. This is why you often see reptiles like lizards or snakes 'sunning themselves'.

Resistance

Electric current travels more easily through a large, thick copper cable than it does through a very thin copper wire.

The thin wire has a **high resistance**. A good conductor with a **low resistance** is a bit like a wide road which can carry a lot of traffic. A wire with a high resistance is like a narrow, bumpy road.

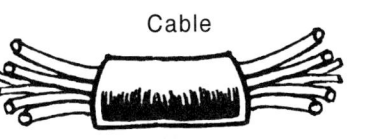

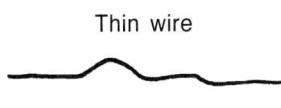

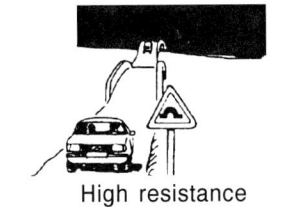

If two bulbs are connected side by side (in series) in a circuit, their resistance adds up and the bulbs are not as bright. If they are connected up in parallel, the bulbs offer less resistance and are lit more brightly.

If two resistances or resistors are joined in **series**, you can find their total resistance by just adding the two up (4 ohms and 4 ohms in series make 8 ohms).

If you join them in **parallel**, the total resistance is half (2 ohms). It is like making the wire twice as thick (or like widening a road to allow less resistance to traffic).

Respiration

All living things get their energy as a result of respiration. There are three stages to respiration in humans. First, breathing brings air into the lungs (or gills for fish). Second, oxygen in the lungs passes into our blood, and carbon dioxide passes out. Third, this oxygen is then used for releasing energy in the cells of our body. Respiration in cells is a bit like burning a fuel. Food combines with oxygen to release energy and make carbon dioxide.

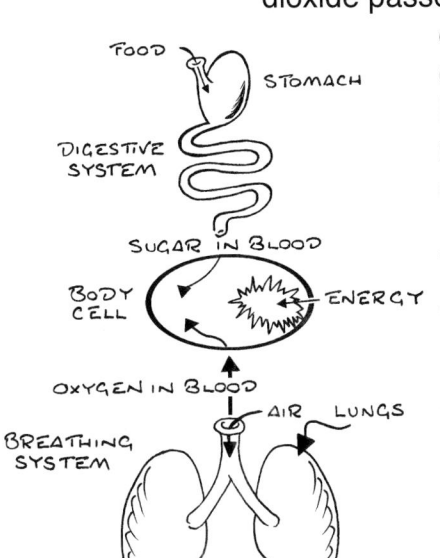

There are two types of respiration: **aerobic** and **anaerobic**. Aerobic respiration happens when there is a good supply of free oxygen, e.g. if you exercise gently, like walking. Anaerobic respiration happens if you do very hard, sudden exercise, e.g. sprinting.

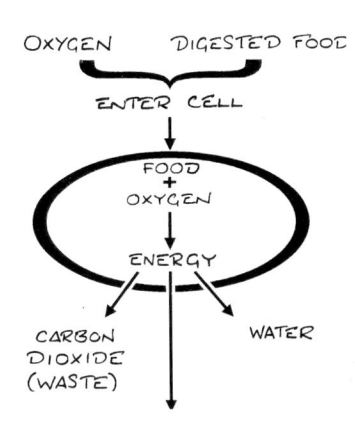

Rocks

There are three types of rock: igneous, sedimentary and metamorphic. **Igneous rocks** are formed from the molten rock or magma inside the Earth.

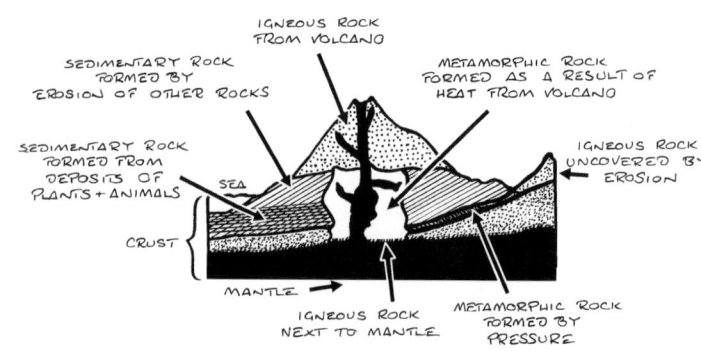

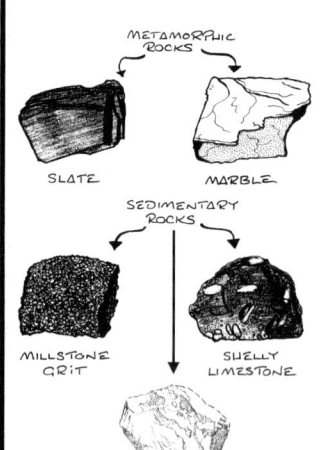

Sedimentary rocks form from the sand, mud and gravel which comes when other rocks are worn away or eroded. Some sediment comes from the remains of dead animals and plants. Layers of sediment build up and are gradually squashed or compressed to make rock, e.g. coal, sandstone, chalk, limestone.

Sedimentary rocks can be changed into much harder rocks by heat or a lot of pressure. This makes them change or metamorphose into new rocks. These are **metamorphic rocks**.

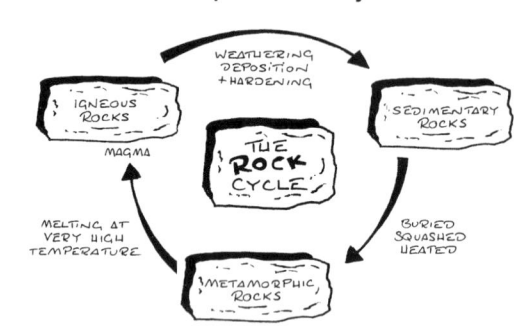

Salt

The salt we put on food is just one type of salt – common salt. But there are lots of other salts – some are used for washing (washing soda), some are bath salts, some relieve tummy trouble.

Most salts are made from acids - the hydrogen atoms in the acid are replaced by a metal atom to make a salt.

Common salt is sodium chloride. Salts can be very useful. Ammonium sulphate is used as a fertiliser. Calcium sulphate is used to make plaster of Paris, which sets hard when it dries – this is used to help heal broken bones.

Rock salt must be cleaned to make the white salt which we use. First it must be washed in lots of water – then the salt dissolves in water but not the sand and bits of rock. These bits are filtered out and then the salt water is heated to leave salt behind.

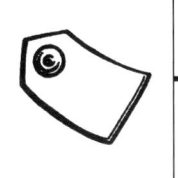

Satellite

An object which goes in orbit around another object is called a satellite. Our Moon is a **natural satellite** of the Earth – it orbits in about 28 days. The first **artificial satellite** was launched by Russia in 1957, SPUTNIK. Nowadays there are hundreds of satellites in circular orbits around the Earth.

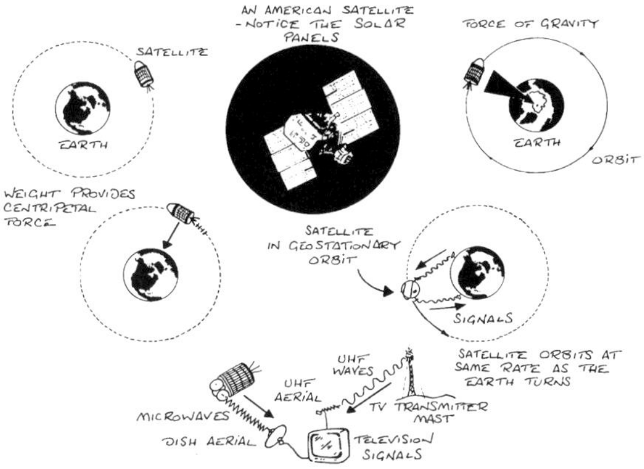

Some are used for communications – they beam microwave signals from one part of the Earth to another. These signals can carry radio or TV programmes, computer data or telephone calls.

Some satellites are used for keeping an eye on the weather. Navigation satellites are used by boats and planes to know their exact position. Satellites have to be launched into space using rockets or be carried up by the Space Shuttle.

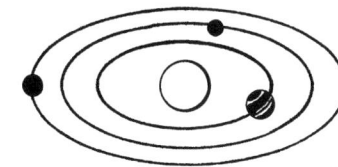

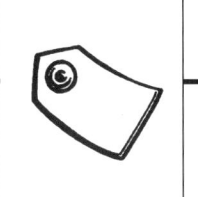

Seasons

Spring, summer, autumn and winter are the four seasons. We have seasons because the Earth is tilted slightly. An imaginary line through the centre of the Earth from North to South Pole is called the **Earth's axis**. In summer in the Northern hemisphere, the axis is tilted towards the Sun. The Earth goes round the Sun in orbit. Six months after midsummer, the Earth's axis is tilted away from the Sun – this is midwinter (the winter **solstice**).

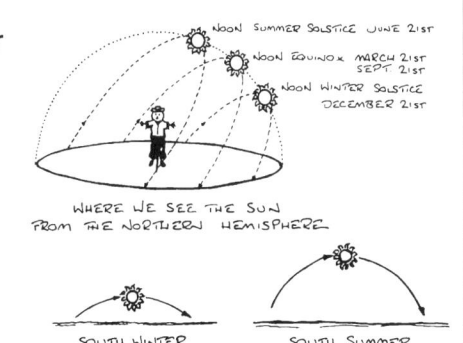

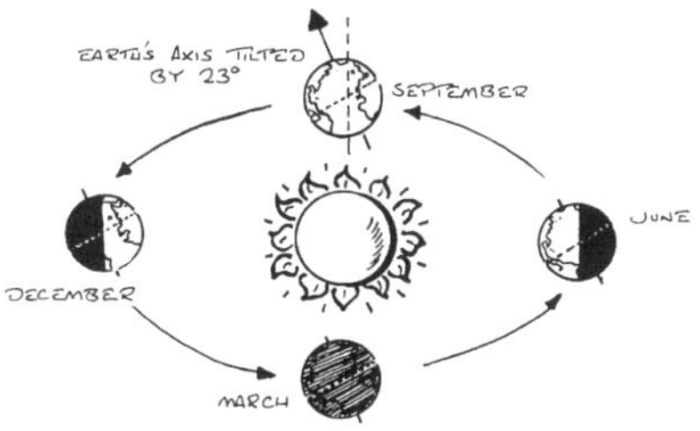

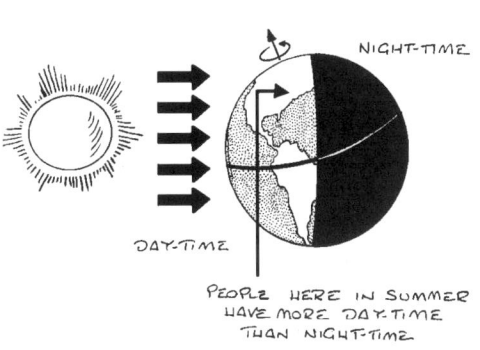

In Summer the Sun gets much higher in the sky – it also rises earlier and sets later. In autumn and spring we have equal times for day and night – the **equinox**.

Series

A row of bulbs connected side by side are in series.

One bulb on its own may be brightly lit using just one cell. But if two bulbs are put together in series, still using one cell, they are only dimly lit. Three bulbs in series with only one cell would be even dimmer. They would need three cells to make them light brightly.

If more and more things are connected in series, their total resistance adds up and more voltage is needed to drive the same current through them. A 4 ohm resistor connected in series with a 5 ohm resistor make a total resistance of 4 + 5 = 9 ohms.

The cells here are connected in series. The total voltage of the 3 cells adds up to 4.5 volts (1.5 + 1.5 + 1.5 = 4.5 volts)

In a series circuit there is only one route that the current can take. If you measure the size of the current with an ammeter at different points in the circuit it will be the same.

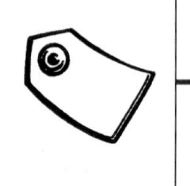

Skeleton

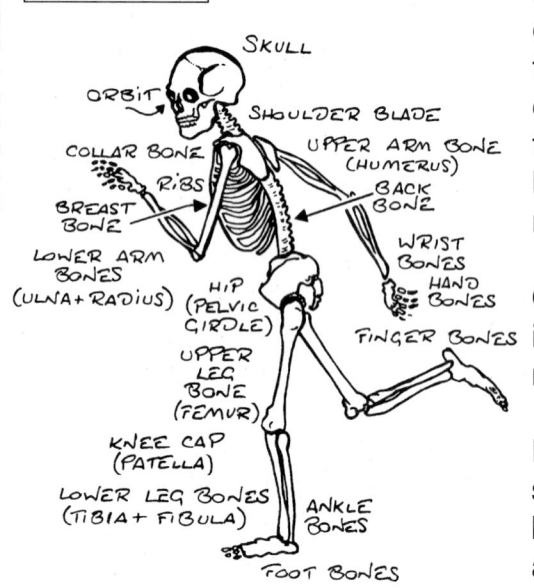

A skeleton is a framework of bones in an animal. The skeleton has four important jobs: to support the body and keep its shape; to allow it to move; to protect vital organs, e.g. the skull protects the brain; to produce blood cells from the marrow in some of the bones. The human skeleton has over 200 bones. Bones are connected at joints like the hip, the elbow or the knee. Bones are held together by ligaments. Bones, muscles and ligaments enable us to move, to chew food, and to push and pull things.

Other animals have skeletons inside their bodies, e.g. fish, reptiles, birds.

But some animals have skeletons outside their bodies, a bit like a suit of armour, e.g. crabs, lobsters, crayfish, insects.

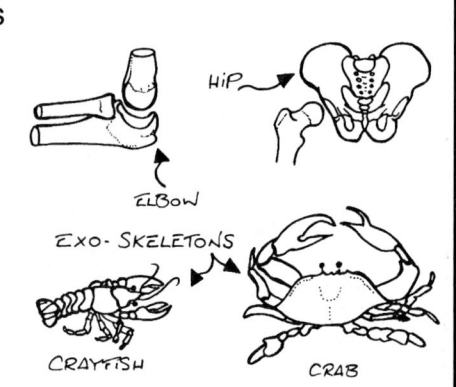

Solid

All the materials on Earth are either solids, liquids or gases.

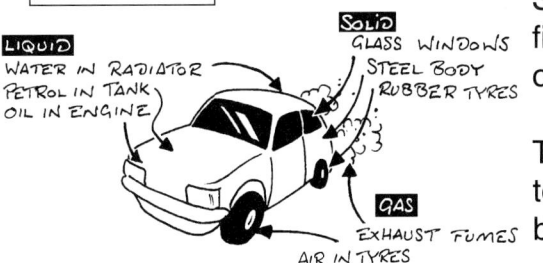

Solids don't need to be kept in a container. They have a fixed shape and volume – unless they are melted down, cut or crushed.

Things can change from one state to another. A solid **melts** to become a liquid. A liquid **boils** or **evaporates** to become a gas.

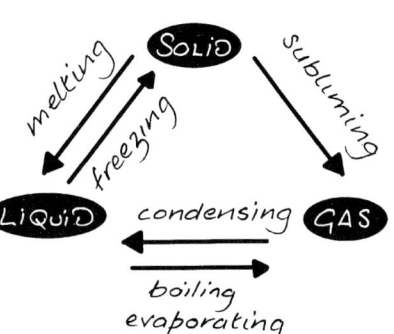

Some materials change directly from a solid to a gas – this is called **subliming**. Carbon dioxide is a solid if it is colder than –55°C. It is then called 'dry ice' because it goes straight into a gas without melting.

Solids, liquids and gases are different because of the way the **particles** inside them are held together and move around.

Soluble

Salt dissolves in water – we say that salt is soluble in water. The solid that dissolves, e.g. salt, is called the **solute**. The liquid it dissolves in is called the **solvent**, e.g. water. When the salt is dissolved in the water, the new mixture (salty water) is called the **solution**.

If you keep trying to dissolve more and more salt in water, there comes a point when no more will dissolve. The solution is then **saturated**. To get more salt to dissolve you would need to warm it up. The Dead Sea is saturated with salt – this makes it very **dense**, and humans can float in it.

Some things don't dissolve in water, but they will dissolve in other liquids. These liquids are called solvents. For example, white spirit is a solvent for some paints. Alcohol is a solvent for some inks and will also dissolve away certain stains (a 'stain remover').

Sound

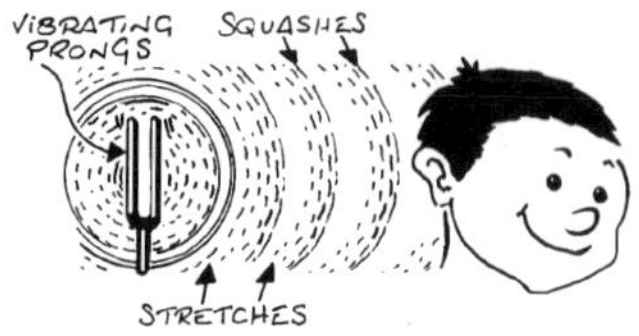

When something moves backwards and forwards, over and over again, it is vibrating. **Vibrations** make sounds. A drum skin, a guitar string and a tuning fork are all vibrating when they make a sound.

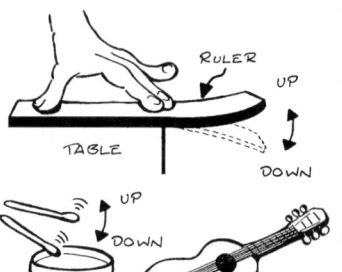

Vibrations make sound waves in the air. The air is 'squashed' and 'stretched' at different points along the wave. If this sound wave reaches your ear, it makes the ear drum vibrate and a message is sent to your brain.

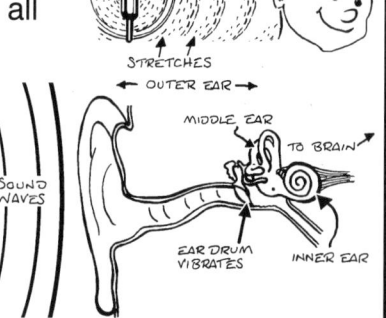

Sound waves always need something to travel in (a 'medium'). Unlike light waves, they cannot travel through empty space. Sound travels at different speeds in different materials.

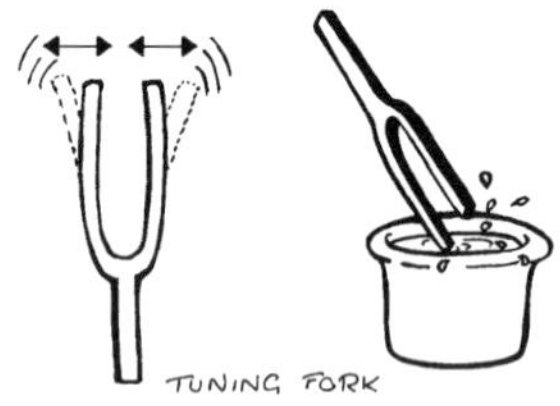

Sounds have different **frequencies** (or pitch). Frequency is measured by the number of vibrations per second using a unit called **hertz** (Hz for short). A very low sound might have a frequency of 50 hertz (50 vibrations per second). A very high note could be 5,000 vibrations per second or 5,000 Hz. This has a very high **pitch**.

Speed

A sprinter runs 100 metres in about 10 seconds. His average speed is 10 metres per second. A fast aeroplane travels about 3000 metres in 10 seconds – its speed is 300 metres per second. A rifle bullet travels about 2700 metres in 3 seconds – its speed is 900 metres per second. To compare the speeds of different moving objects you work out how far they travel in one second.

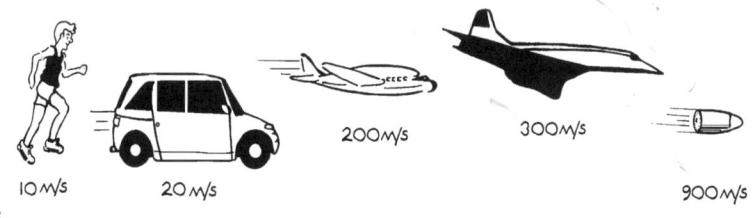

SPEED = $\dfrac{\text{DISTANCE TRAVELLED}}{\text{TIME TAKEN}}$

So the sprinter's speed = $\dfrac{100 \text{ metres}}{10 \text{ seconds}}$ = 10 metres per second

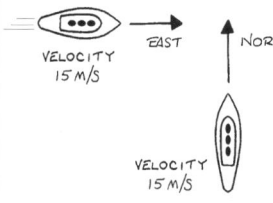

If you know the speed of something, you know how far it travels in one second. But it is often useful to know its direction too. The **velocity** of a moving object tells you its speed and the direction it is moving in, e.g. North-West, or South-East. These two ships on the sea have the same speed of 15 m/s, but their velocity is different – will they collide?

Sperm

Sperms are male sex cells – they look like tiny tadpoles.
In fertilisation, sperms swim towards the **ovum** of a female. If one sperm gets inside, its **nucleus** joins with the female nucleus to make a fertilised ovum. This divides into many cells and goes on to become an **embryo** and eventually a baby.

Human sperms come from the testes of the male – they travel up through a sperm tube (or duct) and out through the penis during intercourse.

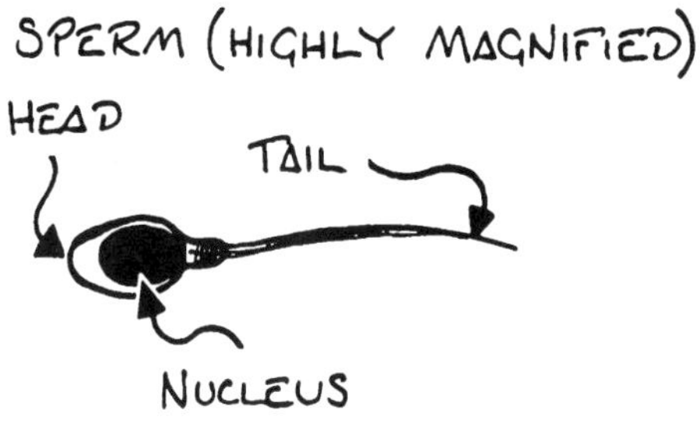

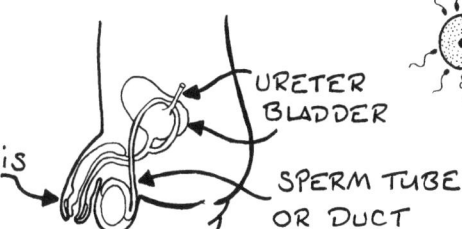

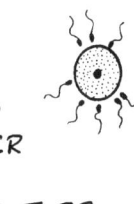

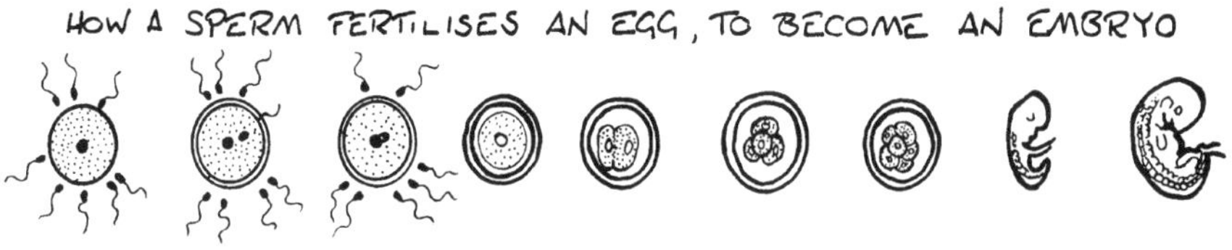

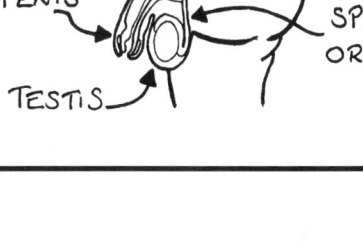

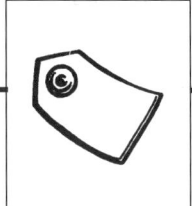

Star

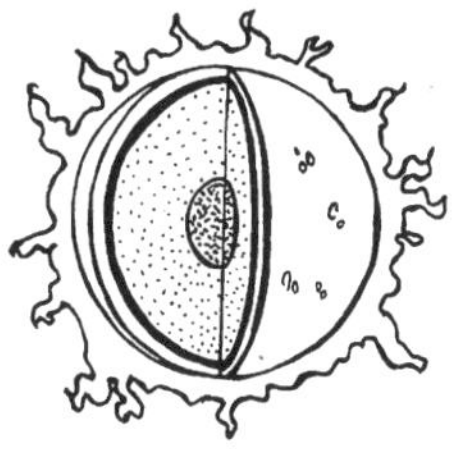

Stars are giant balls of glowing gas. They give off light, infrared (heat) and other waves. Our nearest star is the Sun. Its energy comes from **nuclear fusion** – atoms of hydrogen crashing into each other at high speeds to release nuclear energy.

The next nearest star is called 'Proxima Centauri'. It is a distance of about 4 light years away. This is the distance travelled by light in 4 years (a very long way). Light from our Sun takes about 8 minutes to reach Earth, so it is a lot closer!

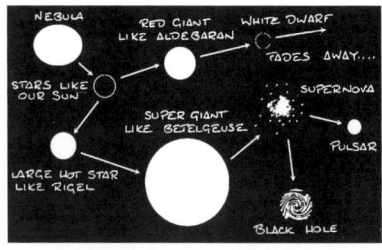

All stars have a lifetime. They are born from a cloud of dust and gas called a **nebula**. This cloud shrinks under the pull of gravity, is squashed into a much smaller volume, and then nuclear reactions start, releasing nuclear energy. The star is born.

There are millions of stars in the sky. When people look at them they seem to form patterns, called **constellations**. Here are two common patterns or constellations seen in the Northern hemisphere – Orion and his belt are seen looking South in Winter. The Plough, which looks like a large saucepan, is seen looking North.

Teeth

There are four types of teeth: sharp **incisors**, used for biting and cutting; pointed **canines**, used for tearing food – animals which hunt and kill have long canines; and flat **pre-molars** and **molars**, used for crushing and grinding.

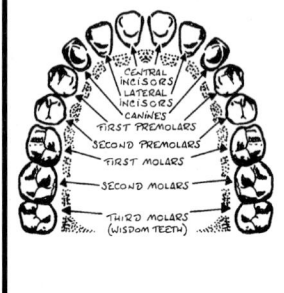

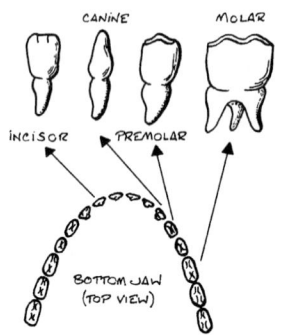

A tooth has several parts: a **crown**, covered by **enamel** to protect it; a **neck**, where it goes into the gums; and the **root**, where it is joined to the jaw bone.

Inside each tooth is a substance like bone, called **dentine** and a soft centre called the **pulp cavity**. This pulp contains the nerves and blood vessels which supply the tooth with food and oxygen.

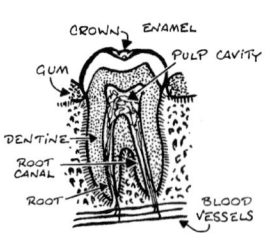

Teeth can decay if old food sticks to them and bacteria start to grow on them. This becomes acid and dissolves the tooth enamel. To prevent tooth decay, you should clean your teeth often, and avoid eating sugary food. Fluoride in toothpaste and drinking water can also help.

Temperature

If you know the temperature of something you know how hot or cold it is on a temperature scale. The most common is the **Celsius scale**. It uses the melting point of pure ice as 0°C, and the boiling point of pure water as 100°C. Here are some other temperatures on the Celsius scale.

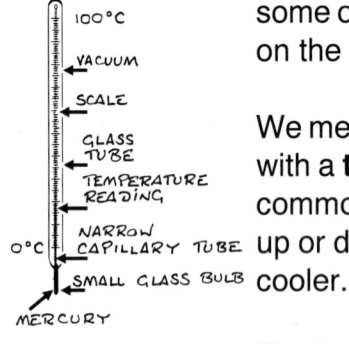

We measure temperature with a **thermometer**. One common type uses a liquid called mercury which goes up or down inside a narrow tube as it gets hotter or cooler.

The human body needs to have a steady temperature of about 37°C. If it gets too hot, sweat is formed on our skin and it cools us down as it evaporates. Our blood vessels (capillaries) also take more blood to the surface of the skin to cool us down. If we get too cool, 'goose pimples' make the hairs on our body stick up and help to trap our body heat in. We also start to shiver which makes us warmer.

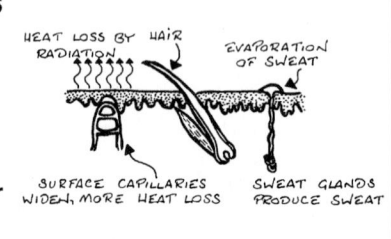

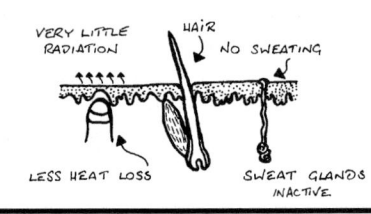

Tendon

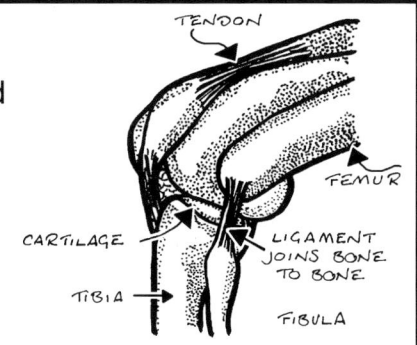

A tendon is a band of tough tissue which connects the end of a muscle to a bone. Tendons are made of strong fibres which hardly stretch at all (unlike a ligament which does stretch – remember: ligaments connect bone to bone).

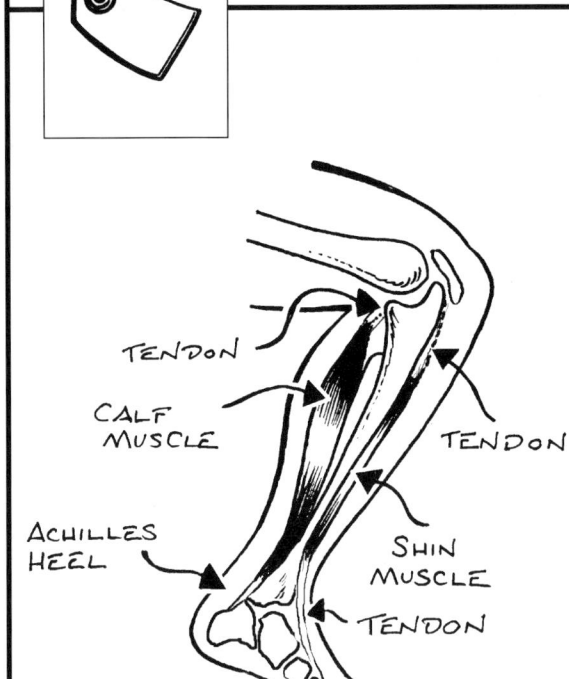

In the leg, tendons connect the thigh muscles and the calf muscles to the knee bones.

At the bottom of the leg, the Achilles tendon connects the calf muscle to the heel bone.

Tendons at the top of your arm join the arm muscles to the shoulder blade.

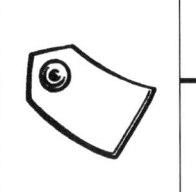

Ultrasonic

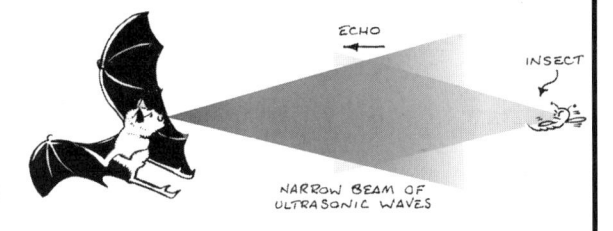

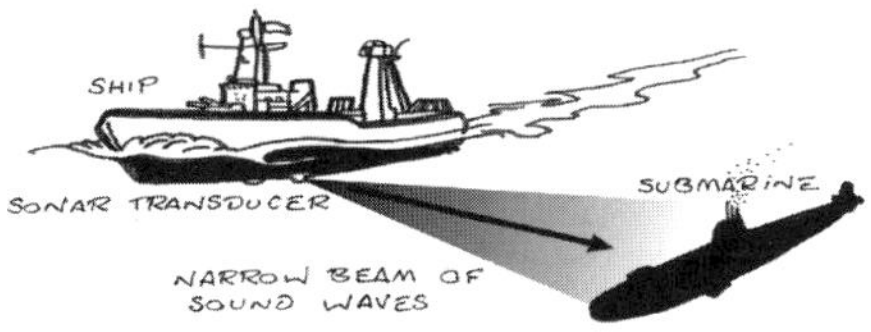

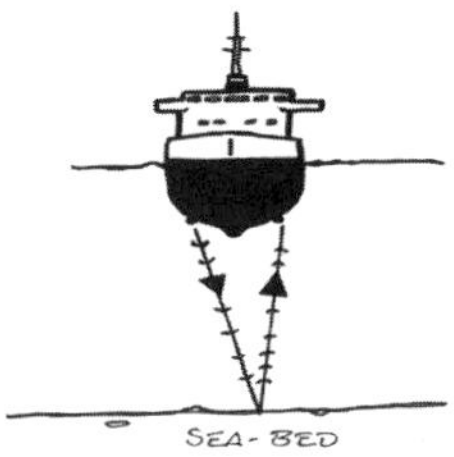

Sounds with a very high frequency which people cannot hear are called ultrasonics or **ultrasounds**. People can hear sounds with a high frequency, up to about 18,000 vibrations per second or 18,000 **hertz**. But many useful sounds have frequencies well above this, 20,000 hertz or higher (up to 150,000 hertz and above). Bats use ultrasonics to 'see' in the dark or spot insects. They send out a beam of ultrasound which bounces back off the insect – the echo tells the bat where the insect is. Dolphins, whales and porpoises use ultrasonic echoes for hunting and navigation. Ships use ultrasounds to see how deep the sea is, or to detect objects underwater.

An ultrasonic beam can also be used to 'see' a baby inside a mother's womb. This is called ultrasonic scanning. Ultrasonic waves can also be used for: cleaning teeth and other objects by literally shaking the dirt off them; or for detecting tiny cracks in metal, concrete or rubber.

Ultraviolet

Ultraviolet waves are the rays from the Sun that give people a suntan. They can also cause skin cancer and damage the retina of the eye. Ultraviolet rays are just to one side of light rays in the family of waves called the electromagnetic spectrum, next to the violet end of the visible spectrum.

They have a shorter wavelength (less than one hundred-millionth of a metre) and a higher frequency than light rays.

Sun-lamps and sunbeds give out ultraviolet rays. These can tan the skin but also damage it and may even cause skin cancer. Special skin creams can stop ultraviolet rays from reaching the skin. The ozone layer high in the atmosphere also filters out some harmful ultraviolet rays, but this layer has been damaged by pollution. This may be one reason why skin cancer is more common now than years ago.

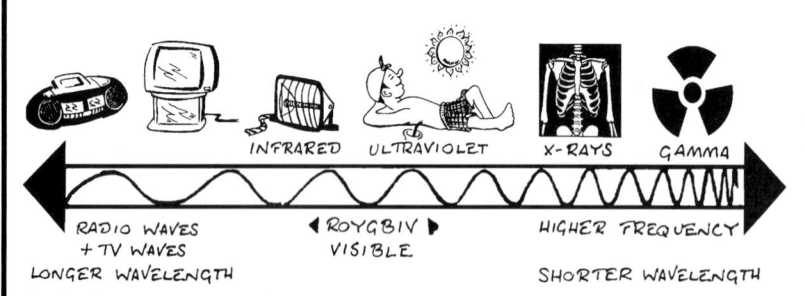

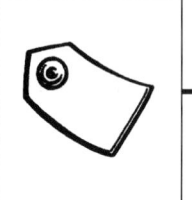

Umbra

Shadows are made when something blocks the light from a luminous object. Shadows form because light travels in straight lines. An **umbra** is a region of very dark shadow, with no light at all in it. A **penumbra** is a region of part shadow. Light coming from a very small bulb makes a dark shadow or umbra when the light is blocked.

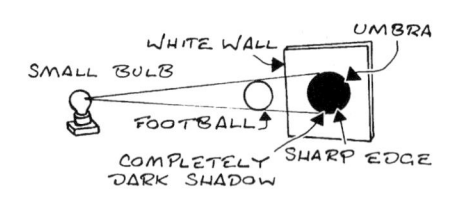

But the light from most bulbs comes from a larger area – when this light meets an object the shadow formed has two parts. The very dark bit is the **umbra**. The region where there is some light (part shadow) is the **penumbra**.

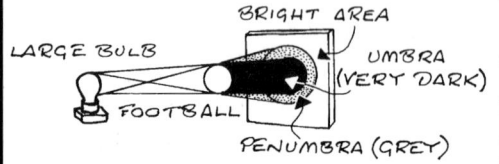

When the Moon casts a shadow on the Earth it forms an umbra and a penumbra.

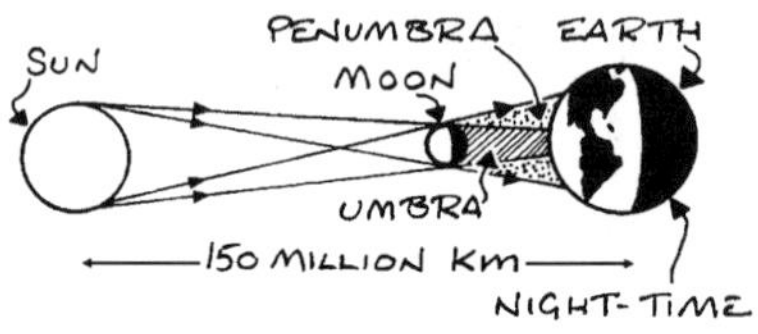

If you are standing in the umbra region you see a **total eclipse**. A person in the penumbra region sees a **partial eclipse**. We sometimes see an **eclipse of the moon** – this happens when the Earth casts a shadow on the surface of the moon.

Urine

Urine is the waste liquid that comes out of your body. The system that makes urine is called the **urinary system**. It is made up of two **kidneys**, two tubes from the kidneys called ureters, the bladder, and another tube called the urethra.

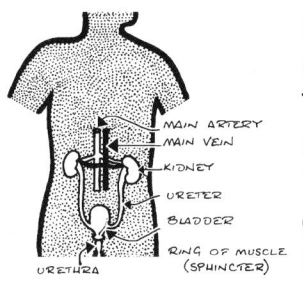

Firstly, the kidneys filter out unwanted substances from the blood. Each kidney removes a waste called urea along with unwanted water and salts. This makes urine which is carried by the tubes called ureters to the bladder. The bladder is a bag which stores urine until it is full and your brain receives a message telling you to go to the toilet. When you go for a wee, a ring of muscle holding the urine in (called the sphincter) is relaxed and out it comes through a tube called the urethra.

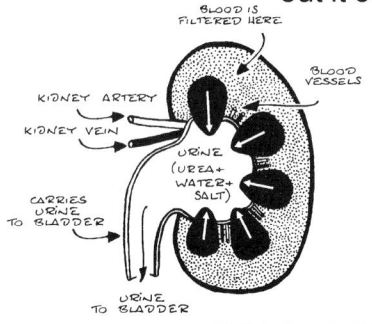

The urinary system can go wrong. If a person suffers from sugar diabetes, then glucose is slowly passed out of the body in the urine. Diabetes can be controlled and treated.

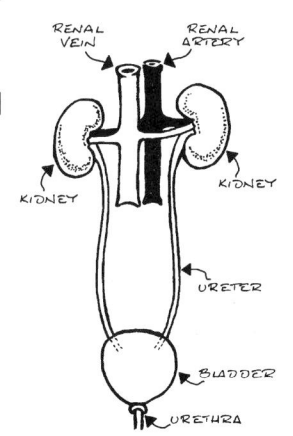

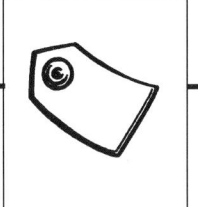

Uterus

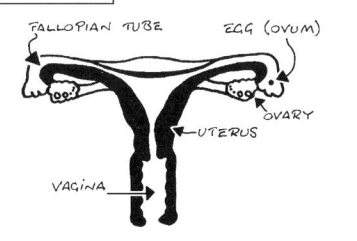

The uterus is more commonly known as the **womb**. This is the place where a baby grows if a mother is pregnant. Each month an egg is produced in the woman's ovary and travels down her fallopian tube to the uterus.

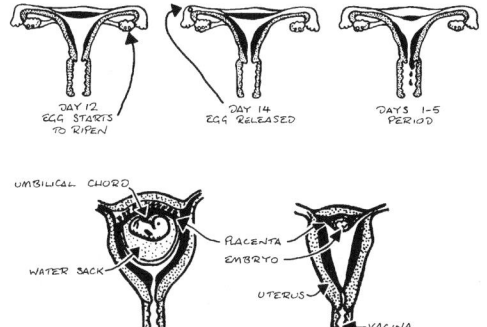

The uterus has a special lining made up of blood vessels, cells and mucus. If the egg is fertilised, this lining is needed to help an **embryo** develop and grow into a baby. If the egg is not fertilised it dies. The inner lining of the uterus is not needed and the body gets rid of it. This is in the form of blood, called a 'period', and is part of the **menstrual cycle**.

Vacuum

A vacuum is a completely empty space. Nobody has ever made a perfect vacuum – there will always be a few air particles or other particles left. The best vacuum we know is in outer space, though even here there may be some particles of gas and dust.

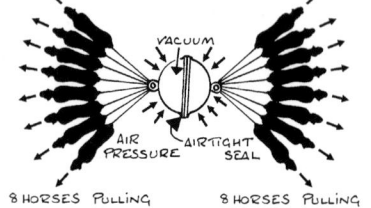

In 1654 the German physicist Otto von Guericke removed the air from inside two metal hemispheres pressed together. Even 16 horses could not pull them apart because of the pressure of the atmosphere forcing them together.

A vacuum is used inside a vacuum flask: there are almost no particles in the vacuum so it does not carry or conduct heat away from the hot liquid inside. Vacuum will not carry sound either. If you place a ringing bell inside a large glass jar and pump the air out, you can still see the bell ringing but you cannot hear it.

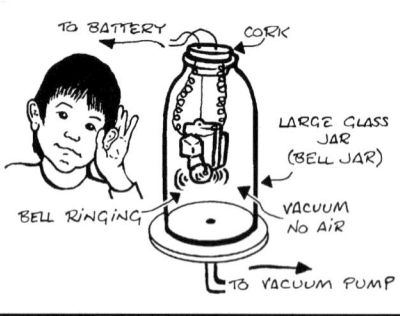

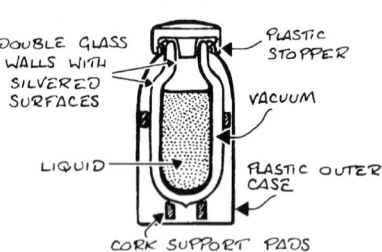

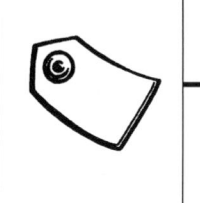

Vertebrate

Animals with backbones are called **vertebrates**. There are five main types of vertebrate. Three types are cold blooded and two types are warm blooded.

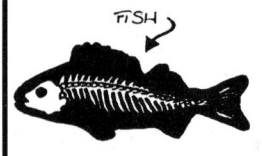

1. **Fish**: these are cold-blooded, live in water, have scales and fins, and breathe through gills.

2. **Amphibians**, such as frogs: these are cold blooded and live in water and on dry land. They lay their eggs in water.

3. **Reptiles**: these are scaly, cold blooded and mostly live on land, although crocodiles live in water. The females lay eggs on land.

4. **Mammals**: these are warm blooded and have fur or hair. Most young mammals are born live (not as eggs) and feed on milk from their mother.

5. **Birds**: these are warm blooded, have feathers and lay eggs.

Virus

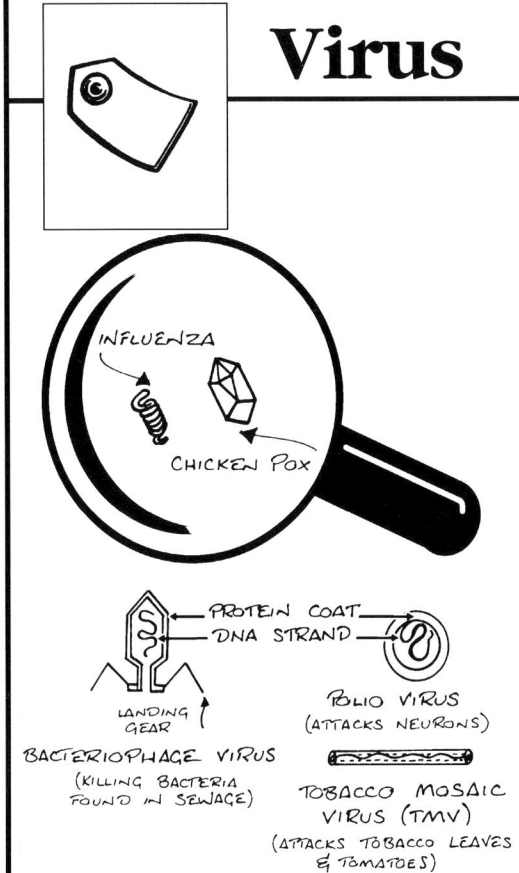

A virus is a tiny bug (microbe) that causes disease in animals or plants. Viruses are much smaller than bacteria. A million of them in a row would only be about 5 mm long.

A virus is a **parasite** – it needs cells from a plant or animal (its host) to live off and reproduce. It takes energy and food from the cells of the host, building hundreds of new viruses and making the animal or plant ill. Viruses have different shapes and sizes.

Scientists are not sure if a virus is a living thing. Some viruses can be dried into a powder and stored for years – yet they become active again when they get on to some living cells. Viruses can't be killed by antibiotics.

Vitamin

Vitamins are chemicals in foods that are essential for the life and health of animals. The body uses them for a lot of different processes. There are 13 main vitamins: A, C, D, E, K, and 8 different B vitamins. Each one is needed for a different job in keeping the body healthy.

We only need vitamins in very small amounts – but we cannot live without them. It is important that vitamins are not destroyed by the ways food is made, or canned, or processed. For example, wholemeal bread is made from whole wheat grain and so it contains all its healthy vitamins and minerals.

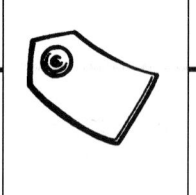

Volume

The volume of something is the amount of space it takes up. Volume is measured in metres cubed (**m³** for short). Smaller volumes (such as the inside of a car engine or motor-bike engine) are measured in centimetres cubed (**cm³**). The volume of liquids is often measured in litres (**l** for short). One litre is the same volume as 1,000 cm³. We also use **millilitres** (ml) for the volume of liquids. There are 1,000ml in 1 litre, so one ml is the same volume as one cm³. You can see the volume of a liquid in a measuring cylinder with a millilitre scale.

If you lower a stone into water inside a measuring cylinder you can see how much the level goes up – the amount the water level goes up by is the volume of the stone.

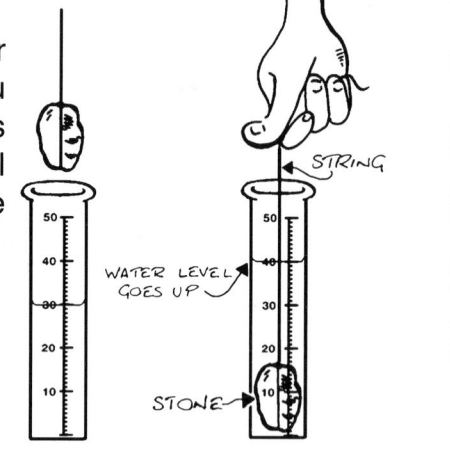

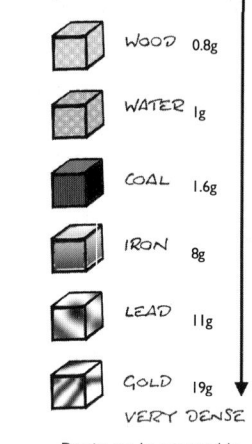

1cm³

NOT SO DENSE
AIR — 0.001g
WOOD — 0.8g
WATER — 1g
COAL — 1.6g
IRON — 8g
LEAD — 11g
GOLD — 19g
VERY DENSE

Density can be compared by comparing the same volume of different materials

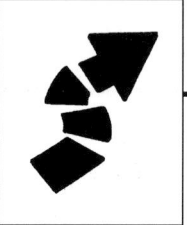

Water/Water cycle

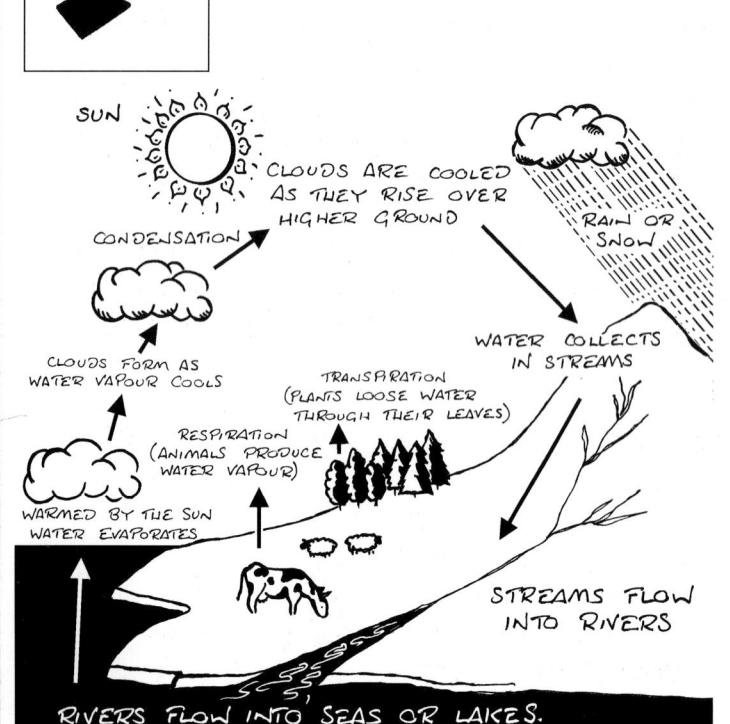

About four-fifths of the Earth's surface is covered by water, in its rivers, seas and lakes. All the time, this water is slowly evaporating into the air to become water vapour. This water vapour rises in the atmosphere, cools down and condenses to make clouds. Eventually, water falls from the clouds as rain, sleet, snow or hail and helps to fill the rivers and lakes. This continuous cycle of water from the Earth's surface up into the clouds, then back down to Earth as rain or snow is called the water cycle.

Water also goes into the air from animals as they breathe out and from plants through their leaves (transpiration).

Watt

Power is measured in watts. The power of something measures how quickly it does work. For example, a large machine or an engine has more power than a human being.

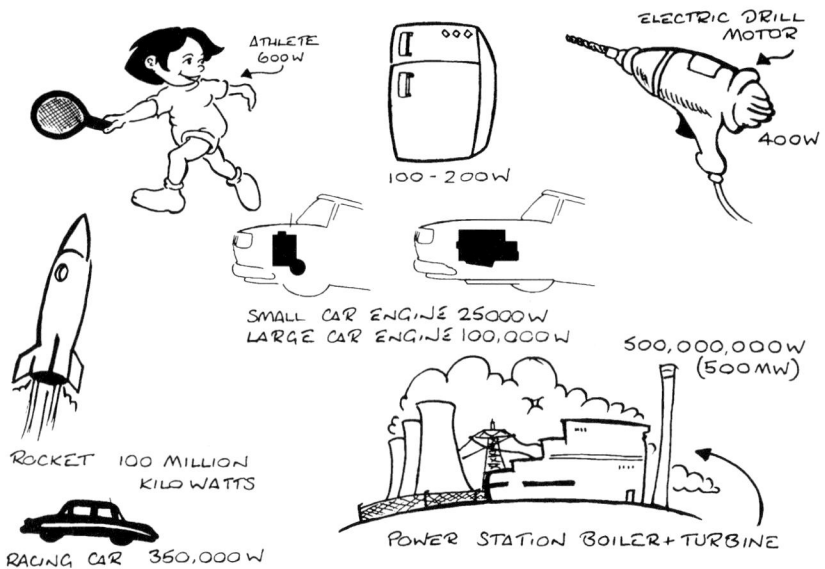

Electrical power is also measured in watts.

One watt is a very small amount of power – so power is often measured in **kilowatts** (kW) which is 1,000 watts. Even larger amounts of power are measured in **megawatts** (MW) which is one million watts. Years ago, power was measured in a unit called horsepower, about 750 watts.

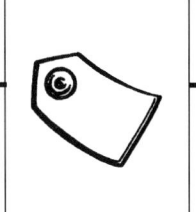

Waves

A wave is a way of carrying energy from one place to another. If you drop a stone in a pond with a ball floating on it, energy from the stone is carried to the ball by water waves or ripples.

There are many different types of wave. Sound is a wave, made by vibrations. Light, heat and X-rays are all waves in a family called the **electromagnetic spectrum**. Waves have some things in common – for example, they are reflected when they reach a barrier. You can show this with water waves in a shallow tank called a ripple tank.

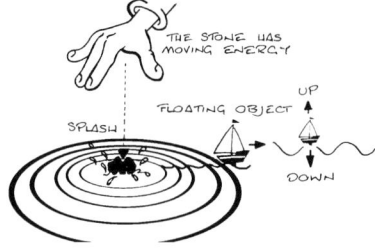

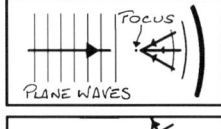

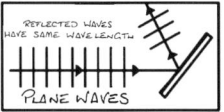

But all waves are different because they have a different **wavelength** and **frequency**. The number of waves passing in one second is called the frequency. The wavelength of a sound wave can be measured in centimetres; medium length radio waves have a wavelength of a few hundred metres; while light waves have a wavelength of less than a millionth of a metre. If you show the wave as a graph, the wavelength is the distance from one peak to the next.

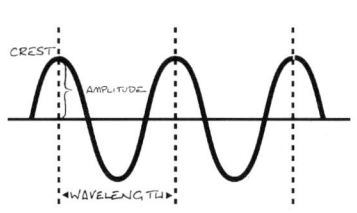

Weight

Weight is the pull of gravity on an object. Here are the weights of some objects on Earth, all measured in **newtons** (N for short):

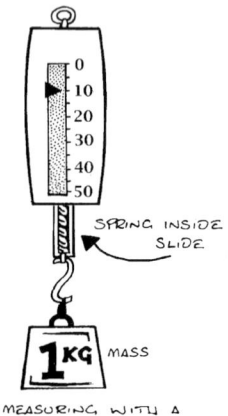

Weight can be measured with a newton meter. This uses a spring which gets stretched more and more as the weight on it gets bigger:

Bag of cement – 500N
World's largest diamond – 6N
Car – 10,000N

Your weight on Earth is the force of attraction between you and the Earth. As you get further and further from the Earth this force gets smaller. A mass of one kilogram (1 kg) has a weight of 10 N on Earth, but 20,000 km away its weight will only be one newton.

A person on the Moon has far less weight than on Earth, because the pull of the Moon's gravity is much less. In outer space, a person far away from any moons or planets might have no weight at all.

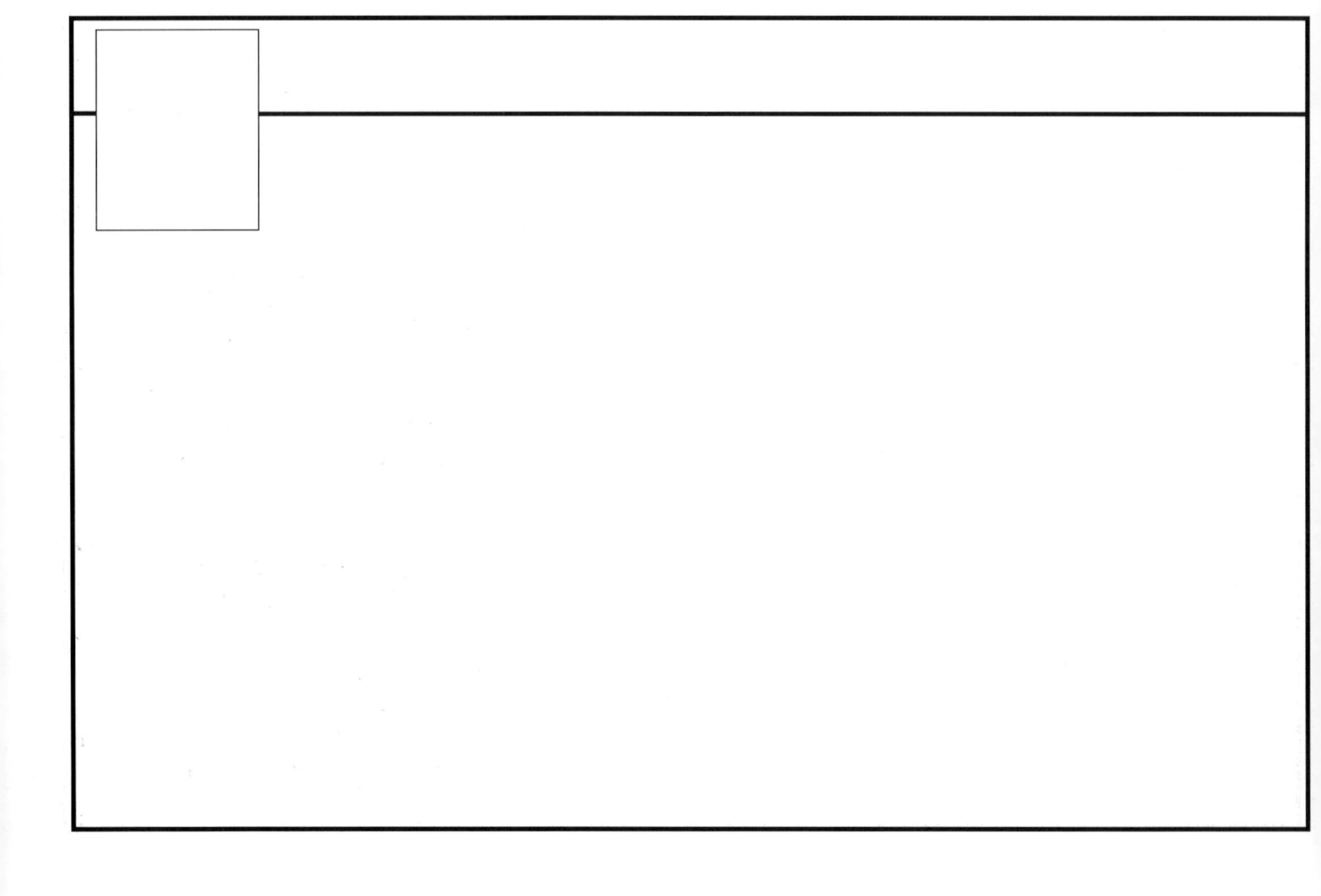

ON EARTH 600 N
ON THE MOON 100 N
IN OUTER SPACE 0 N

MEASURING WITH A NEWTON-METER
SPRING INSIDE SLIDE
1 KG MASS

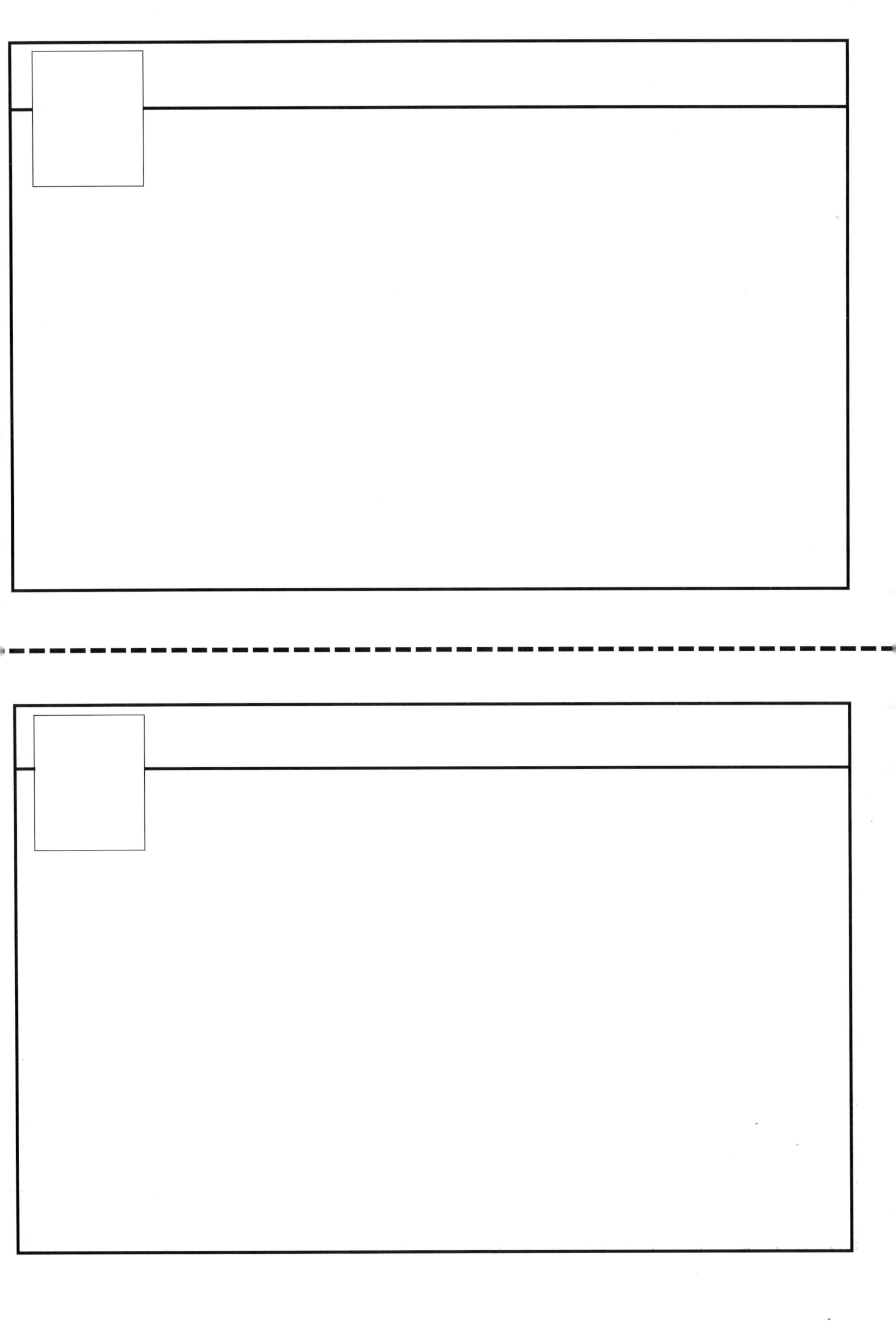

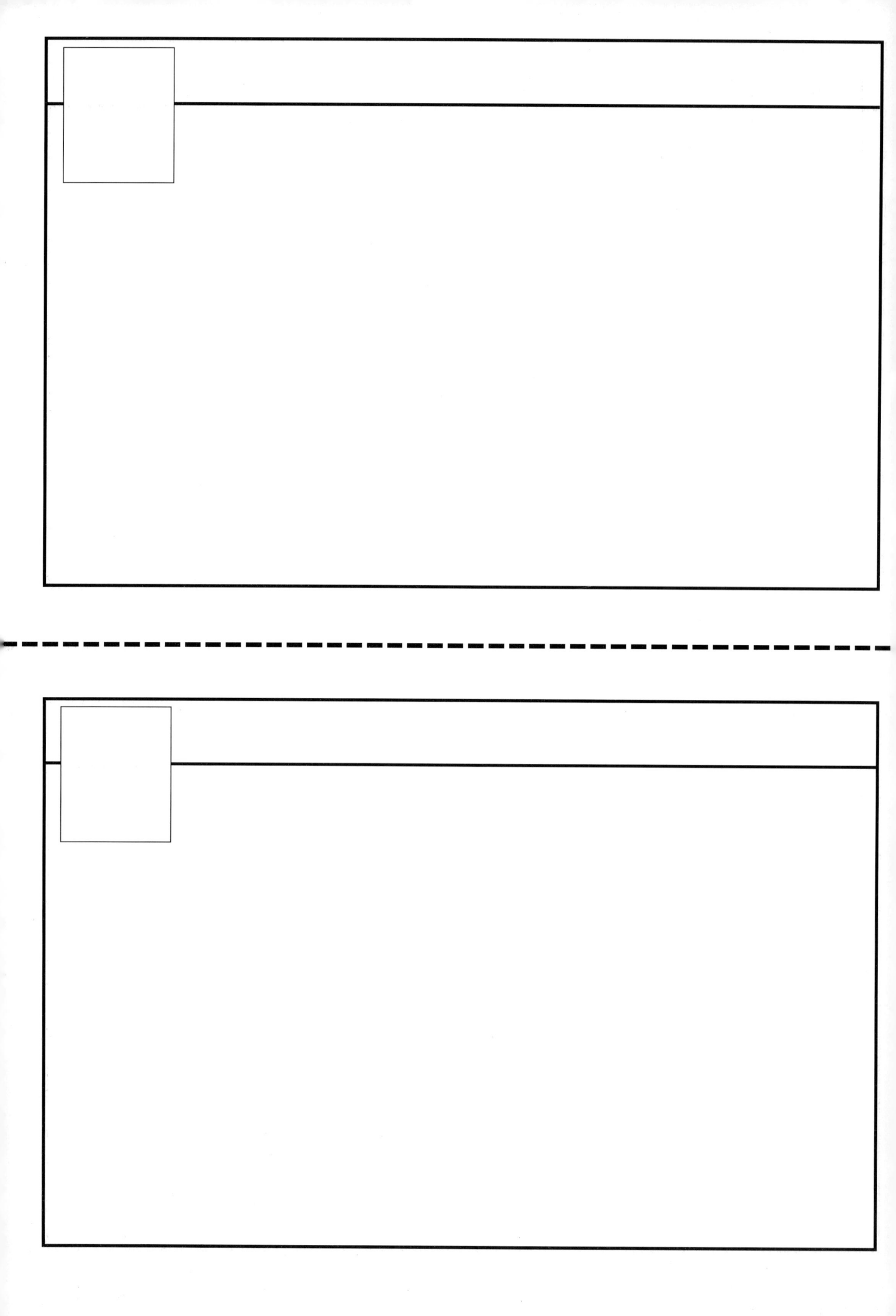

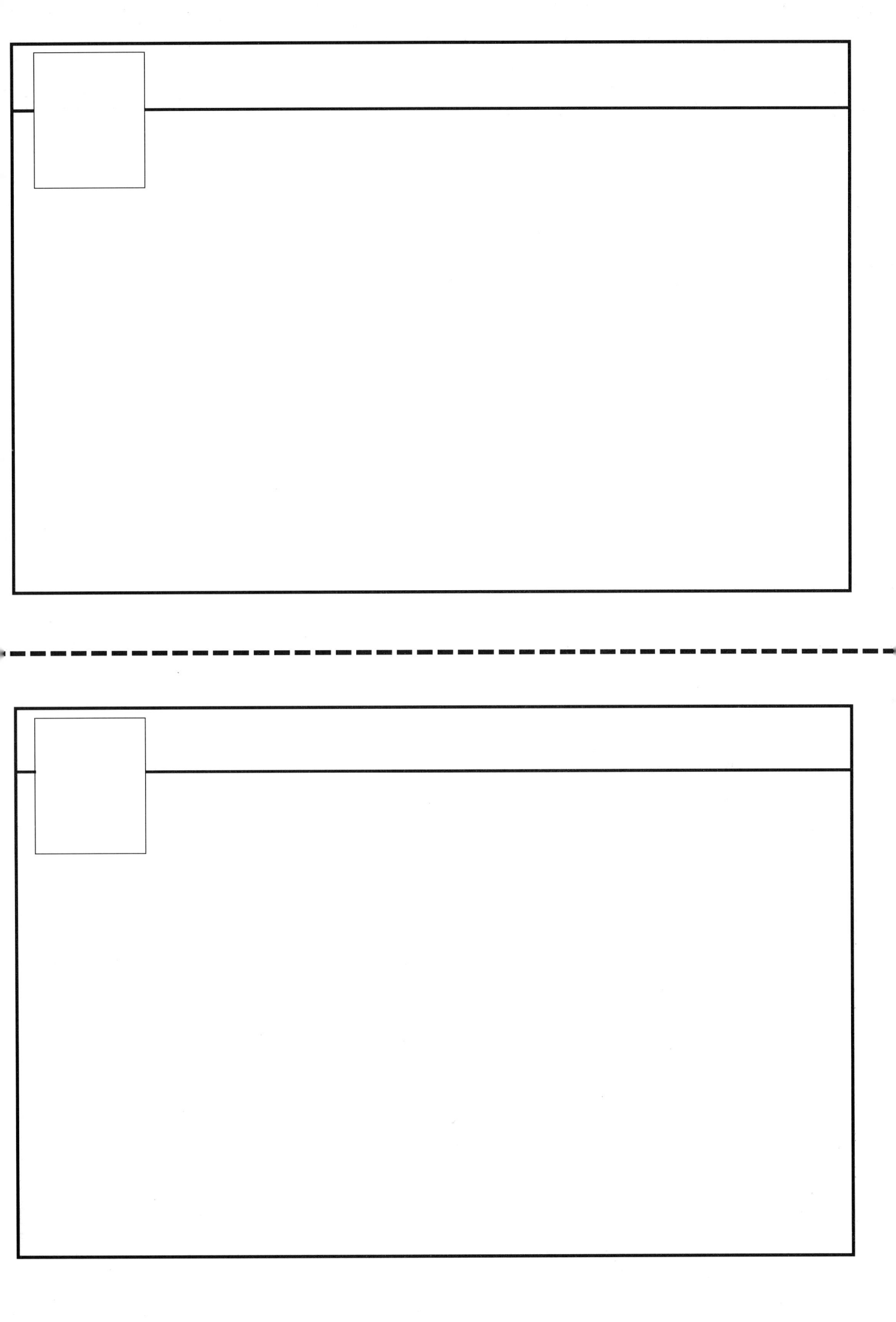